Molecular Mechanisms of Tumor Progression and New Therapeutic Strategies for Urological Cancers, 2nd Edition

Molecular Mechanisms of Tumor Progression and New Therapeutic Strategies for Urological Cancers, 2nd Edition

Guest Editors

Vicenç Ruiz de Porras

Albert Font

MDPI

Basel • Beijing • Wuhan • Barcelona • Belgrade • Novi Sad • Cluj • Manchester

Guest Editors

Vicenç Ruiz de Porras
Department of Pharmacology, Toxicology and Therapeutic Chemistry
University of Barcelona
Barcelona
Spain

Albert Font
Badalona–Badalona Applied Research Group in Oncology (B·ARGO)
Catalan Institute of Oncology
Barcelona
Spain

Editorial Office
MDPI AG
Grosspeteranlage 5
4052 Basel, Switzerland

This is a reprint of the Special Issue, published open access by the journal *International Journal of Molecular Sciences* (ISSN 1422-0067), freely accessible at: https://www.mdpi.com/journal/ijms/special_issues/59I050Q171.

For citation purposes, cite each article independently as indicated on the article page online and as indicated below:

Lastname, A.A.; Lastname, B.B. Article Title. *Journal Name* **Year**, *Volume Number*, Page Range.

ISBN 978-3-7258-6646-5 (Hbk)
ISBN 978-3-7258-6647-2 (PDF)
https://doi.org/10.3390/books978-3-7258-6647-2

Contents

About the Editors

Vicenç Ruiz de Porras

Vicenç Ruiz de Porras is a biologist specialized in Biomedicine (University of Barcelona, 2009) with over fifteen years of experience in translational oncology, biomarker discovery, and academic teaching. He holds an MSc in Applied Clinical Research in Health Sciences (2012) and a PhD with honors in Biomedicine (Autonomous University of Barcelona, 2016). His doctoral and postdoctoral research at the Germans Trias i Pujol Research Institute (IGTP) focused on chemoresistance mechanisms and predictive biomarkers in advanced colorectal and urological cancers. Dr. Ruiz de Porras has co-led competitive national and industry-funded projects as co-principal investigator, establishing new research lines on treatment resistance and therapeutic combinations in metastatic prostate and bladder cancer. He has authored numerous original research and review articles—most of them published in leading Q1 journals—and has contributed to congress presentations, invited talks, and book chapters. In parallel, he has developed extensive teaching and academic experience at the University of Barcelona. Since June 2025, he has been serving as Assistant Professor in the Unit of Toxicology within the Department of Pharmacology, Toxicology, and Therapeutic Chemistry at the Faculty of Pharmacy and Food Sciences, where he combines teaching with research on the biological mechanisms underlying chemotherapy-induced toxicity and the potential use of natural compounds, particularly polyphenols, to mitigate these effects. His academic and research interests focus on biomarker-driven therapy, mechanisms of treatment resistance, and the development of translational strategies to enhance therapeutic efficacy and safety in urological and digestive cancers.

Albert Font

Albert Font holds a degree in Medicine and Surgery from the University of Barcelona and a PhD in Medicine from the Autonomous University of Barcelona. He is currently Head of the Urological Tumours Unit at the Medical Oncology Service of the Catalan Institute of Oncology (ICO) at Germans Trias i Pujol Hospital in Badalona, and serves as the ICO cross-sectional reference for prostate cancer. As coordinator of the Functional Units at HUGTiP and ICO Badalona, he played a key role in establishing the Urological Tumours Functional Unit, which has become a cornerstone for clinical and translational research in urological cancers at the institution. His clinical and research activity focuses on the treatment of urological malignancies, particularly bladder cancer and metastatic prostate cancer. Dr. Font Pous has participated as principal investigator in numerous international phase I, II, and III clinical trials aimed at developing new therapeutic strategies, both within national and international consortia. Notable contributions include studies published in *Nature Medicine*, *The Lancet Oncology*, *European Urology*, and *European Journal of Cancer*. Dr. Font Pous has also been actively engaged in translational research collaborations with international partners, exploring molecular biomarkers of response to platinum-based chemotherapy and the role of neoadjuvant treatment in muscle-invasive bladder cancer. His current research focuses on the molecular mechanisms underlying chemoresistance in prostate and bladder cancer, aiming to identify predictive biomarkers and novel therapeutic targets.

Preface

This Reprint compiles a collection of original research articles and reviews published in the Special Issue Molecular Mechanisms of Tumor Progression and New Therapeutic Strategies for Urological Cancers, 2nd Edition, originally featured in the *International Journal of Molecular Sciences*. The contributions included here provide an updated and comprehensive overview of the biological processes that drive tumor initiation, progression, and therapeutic resistance in urological malignancies, with a particular focus on prostate, bladder, and kidney cancers.

The aim of this Reprint is to offer readers an integrated understanding of the molecular pathways involved in tumor evolution and to highlight innovative diagnostic and therapeutic strategies that are currently shaping translational research in the field. By bringing together studies that span from molecular and cellular biology to clinical applications, this collection intends to serve as a bridge between basic science and clinical oncology.

The motivation behind this work arises from the rapid advances in molecular technologies and precision medicine, which are redefining how we classify and treat urological cancers. Understanding the intricate molecular landscape of these tumors is essential for the development of more effective, personalized therapeutic approaches and for improving patient outcomes.

This Reprint is addressed to researchers, clinicians, and students with an interest in cancer biology, molecular pathology, and translational oncology. We hope that the selected contributions will stimulate further investigation, foster collaboration across disciplines, and inspire new perspectives in the study and management of urological cancers.

Vicenç Ruiz de Porras and Albert Font
Guest Editors

International Journal of
Molecular Sciences

MDPI

Editorial

Molecular Mechanisms of Tumor Progression and New Therapeutic Strategies for Urological Cancers, 2nd Edition

Vicenç Ruiz de Porras [1,2,3]

[1] GRET and Toxicology Unit, Department of Pharmacology, Toxicology and Therapeutic Chemistry, Faculty of Pharmacy and Food Sciences, University of Barcelona, 08028 Barcelona, Spain; vruizdeporras@ub.edu; Tel.: +34-606587010
[2] Badalona Applied Research Group in Oncology (B·ARGO), Catalan Institute of Oncology, Camí de les Escoles, s/n, 08916 Badalona, Spain
[3] CARE Program, Germans Trias i Pujol Research Institute (IGTP), Camí de les Escoles, s/n, 08916 Badalona, Spain

Urological cancers, including malignancies of the bladder, kidney, prostate, upper urinary tract, and penis, are among the most common cancers globally [1] and present significant heterogeneity in terms of clinical behavior, molecular biology, and therapeutic responsiveness [2]. Despite considerable advancements in local and systemic treatments, including immunotherapy and targeted therapies, the prognosis for many patients with advanced disease remains poor. Understanding the underlying molecular mechanisms of tumor progression and treatment resistance is essential to improve existing strategies and identify novel therapeutic targets.

In this second edition of the Special Issue "Molecular Mechanisms of Tumor Progression and New Therapeutic Strategies for Urological Cancers," we feature seven contributions that shed light on critical aspects of tumor biology, biomarker discovery, and therapeutic innovation across various urological malignancies. These studies provide compelling evidence of the evolving landscape in precision oncology and liquid biopsy applications.

Non-invasive biomarkers based on liquid biopsy, capable of dynamically reflecting tumor behavior, are critical for real-time monitoring and therapeutic decision making [3]. In this context, two studies in the current Special Issue focus on circulating tumor-derived components as emerging tools in the management of urological tumors.

Carrasco et al. [4] explored a tumor-agnostic circulating tumor DNA (ctDNA) approach for monitoring patients with muscle-invasive bladder cancer (MIBC) after radical cystectomy (RC). By targeting common mutations (TERT and ATM) using droplet digital PCR, they demonstrated that ctDNA positivity was significantly associated with tumor progression at multiple timepoints. Importantly, ctDNA dynamics provided prognostic information superior to conventional imaging, aligning with prior reports that underscore the utility of ctDNA for early relapse detection [5,6] as well as for predicting immunotherapy efficacy in the adjuvant setting of MIBC [7]. This reinforces ctDNA as a minimally invasive and cost-effective strategy that can overcome the shortcomings of tissue-based biomarkers—including limited tissue availability, reliance on archival samples, and tumor heterogeneity—and is, as a result, increasingly used to guide treatment decisions [8].

In addition, Figols and colleagues [9] reviewed the role of tumor-educated platelets (TEPs) as a liquid biopsy source across urological tumors. TEPs, altered by interactions with tumors and their microenvironment, carry transcriptomic and proteomic signatures reflective of tumor burden and biology [10,11]. Their analysis highlighted TEPs' diagnostic potential, especially in early cancer detection and non-invasive therapy monitoring, and support recent findings in other solid tumors, including non-small cell lung, glioblastoma, and pancreatic cancer [10,12,13].

On the other hand, treatment resistance remains a major challenge, particularly in bladder cancer, where cisplatin-based chemotherapy is still a cornerstone despite limited response rates [14]. Therefore, a better understanding of the mechanisms underlying therapy resistance is essential to developing novel treatment strategies and predicting drug response, ultimately advancing progress toward precision medicine. In this context, two articles in this Special Issue contribute to unraveling the molecular resistance mechanisms.

Elahi Najafi et al. [15] identified GABBR2 as a novel downstream effector of androgen receptor (AR) signaling, conferring cisplatin resistance in bladder cancer cells. The authors demonstrated that AR activation upregulates GABBR2, while its inhibition sensitizes cells to cisplatin. These findings suggest that targeting the AR-GABBR2 axis could reverse resistance, especially in AR-positive tumors. This is consistent with previous evidence that androgen signaling modulates bladder tumorigenesis, tumor progression, and therapy response, suggesting a potential role for androgen deprivation therapies in patients with bladder cancer [16,17].

In a related investigation, Vukovic et al. [18] explored the prognostic relevance of the IL-1β/IL-1RA axis in invasive bladder cancer. Their immunohistochemical analysis revealed that IL-1β expression correlated with aggressive pathological features but paradoxically conferred favorable survival. Moreover, IL-1β levels were positively associated with AKT and Ki-67 expression, implicating a role in autophagy and proliferation. These results reflect the complex dual role of inflammation in cancer, where pro-inflammatory cytokines may drive both tumor progression and anti-tumor immune responses [19].

In clear-cell renal cell carcinoma (ccRCC), extracellular matrix remodeling enzymes contribute to angiogenesis, invasiveness, and metastasis [20]. Młynarczyk et al. [21] compared MMP-14 and MMP-15 expression and activity in tumor and normal kidney tissues. They found that MMP-14 was significantly more active and expressed, particularly in higher-grade tumors, supporting its role in angiogenesis and invasiveness. These results reinforce earlier reports that link MMP-14 to disease progression and poor short-term prognosis in RCC and bladder cancer [22,23].

Although rare, penile squamous cell carcinoma (SCC) presents therapeutic challenges in advanced stages. Fahey et al. [24] described a case of metastatic penile SCC with a durable response to the antibody–drug conjugate, enfortumab vedotin, a human anti-nectin-4 antibody linked to the cytotoxic microtubule-disrupting agent monomethyl auristatin E [25]. Based on the results of the EV-301 phase III trial [26], enfortumab vedotin was approved by the FDA in 2019 for the treatment of patients with locally advanced or metastatic urothelial carcinoma who were previously treated with platinum-based chemotherapy and checkpoint inhibitor therapy. Interestingly, this report highlights the potential application of novel targeted agents beyond their initial indications and supports further investigation into Nectin-4 as a therapeutic target in penile cancer, consistent with expression patterns reported in urothelial and squamous histologies [25,27].

Prostate cancer (PCa), especially in its metastatic castration-resistant form (mCRPC), remains a challenging clinical entity that, despite significant advances in the treatment in the last decade, still carries a poor prognosis, with a median overall survival of approximately three years [28]. In this context, Porcel-Pastrana and colleagues [19] explored a combinatorial approach using hydroxytyrosol (HT), a phenolic compound from olive oil, and metformin, a widely used antidiabetic agent. Their in vitro data showed a synergistic inhibition of cell proliferation, migration, and tumorsphere formation across multiple

PCa cell lines. Mechanistically, the combination suppressed key oncogenic pathways including MAPK, AKT, and TGF-β. These findings build upon prior preclinical evidence supporting the repurposing of metabolic regulators—primarily drugs used to treat diabetes, such as metformin, or to lower cholesterol, such as simvastatin—for cancer treatment and prevention [29–31].

Collectively, the contributions in this Special Issue reflect the diversity and depth of current research in urological oncology. From liquid biopsy technologies to resistance mechanisms and novel therapeutics, these studies point toward a future where disease management is increasingly personalized, biology-driven, and minimally invasive. We hope this compilation stimulates further research and translational efforts, ultimately improving outcomes for patients affected by these complex malignancies.

Acknowledgments: The authors thank Sara Bystrup for assistance in drafting the manuscript.

Conflicts of Interest: The author declares no conflicts of interest.

References

1. Bray, F.; Laversanne, M.; Sung, H.; Ferlay, J.; Siegel, R.L.; Soerjomataram, I.; Jemal, A. Global cancer statistics 2022: GLOBOCAN estimates of incidence and mortality worldwide for 36 cancers in 185 countries. *CA Cancer J. Clin.* **2024**, *74*, 229–263. [CrossRef]
2. Antar, R.M.; Fawaz, C.; Gonzalez, D.; Xu, V.E.; Drouaud, A.P.; Krastein, J.; Pio, F.; Murdock, A.; Youssef, K.; Sobol, S.; et al. The Evolving Molecular Landscape and Actionable Alterations in Urologic Cancers. *Curr. Oncol.* **2024**, *31*, 6909–6937. [CrossRef]
3. Ma, L.; Guo, H.; Zhao, Y.; Liu, Z.; Wang, C.; Bu, J.; Sun, T.; Wei, J. Liquid biopsy in cancer current: Status, challenges and future prospects. *Signal Transduct. Target. Ther.* **2024**, *9*, 336. [CrossRef] [PubMed]
4. Carrasco, R.; Ingelmo-Torres, M.; Trullas, R.; Roldan, F.L.; Rodriguez-Carunchio, L.; Juez, L.; Sureda, J.; Alcaraz, A.; Mengual, L.; Izquierdo, L. Tumor-Agnostic Circulating Tumor DNA Testing for Monitoring Muscle-Invasive Bladder Cancer. *Int. J. Mol. Sci.* **2023**, *24*, 16578. [CrossRef]
5. Christensen, E.; Birkenkamp-Demtroder, K.; Sethi, H.; Shchegrova, S.; Salari, R.; Nordentoft, I.; Wu, H.T.; Knudsen, M.; Lamy, P.; Lindskrog, S.V.; et al. Early Detection of Metastatic Relapse and Monitoring of Therapeutic Efficacy by Ultra-Deep Sequencing of Plasma Cell-Free DNA in Patients with Urothelial Bladder Carcinoma. *J. Clin. Oncol.* **2019**, *37*, 1547–1557. [CrossRef]
6. Christensen, E.; Nordentoft, I.; Birkenkamp-Demtroder, K.; Elbaek, S.K.; Lindskrog, S.V.; Taber, A.; Andreasen, T.G.; Strandgaard, T.; Knudsen, M.; Lamy, P.; et al. Cell-Free Urine and Plasma DNA Mutational Analysis Predicts Neoadjuvant Chemotherapy Response and Outcome in Patients with Muscle-Invasive Bladder Cancer. *Clin. Cancer Res.* **2023**, *29*, 1582–1591. [CrossRef]
7. Powles, T.; Assaf, Z.J.; Davarpanah, N.; Banchereau, R.; Szabados, B.E.; Yuen, K.C.; Grivas, P.; Hussain, M.; Oudard, S.; Gschwend, J.E.; et al. ctDNA guiding adjuvant immunotherapy in urothelial carcinoma. *Nature* **2021**, *595*, 432–437. [CrossRef]
8. Corcoran, R.B.; Chabner, B.A. Application of Cell-free DNA Analysis to Cancer Treatment. *N. Engl. J. Med.* **2018**, *379*, 1754–1765. [CrossRef] [PubMed]
9. Figols, M.; Chekhun, S.; Fernandez-Saorin, M.; Perez-Criado, I.; Bautista, A.; Font, A.; Ruiz de Porras, V. Tumor-Educated Platelets in Urological Tumors: A Novel Biosource in Liquid Biopsy. *Int. J. Mol. Sci.* **2025**, *26*, 3595. [CrossRef] [PubMed]
10. Best, M.G.; Sol, N.; Kooi, I.; Tannous, J.; Westerman, B.A.; Rustenburg, F.; Schellen, P.; Verschueren, H.; Post, E.; Koster, J.; et al. RNA-Seq of Tumor-Educated Platelets Enables Blood-Based Pan-Cancer, Multiclass, and Molecular Pathway Cancer Diagnostics. *Cancer Cell* **2015**, *28*, 666–676. [CrossRef]
11. Best, M.G.; Wesseling, P.; Wurdinger, T. Tumor-Educated Platelets as a Noninvasive Biomarker Source for Cancer Detection and Progression Monitoring. *Cancer Res.* **2018**, *78*, 3407–3412. [CrossRef]
12. Nilsson, R.J.; Balaj, L.; Hulleman, E.; van Rijn, S.; Pegtel, D.M.; Walraven, M.; Widmark, A.; Gerritsen, W.R.; Verheul, H.M.; Vandertop, W.P.; et al. Blood platelets contain tumor-derived RNA biomarkers. *Blood* **2011**, *118*, 3680–3683. [CrossRef] [PubMed]
13. Zaki-Dizaji, M.; Taheri, Z.; Heiat, M.; Hushmandi, K. Tumor-educated platelet, a potential liquid biopsy biosource in pancreatic cancer: A review. *Pathol. Res. Pract.* **2025**, *270*, 155986. [CrossRef] [PubMed]
14. Ruiz de Porras, V.; Pardo, J.C.; Etxaniz, O.; Font, A. Neoadjuvant therapy for muscle-invasive bladder cancer: Current clinical scenario, future perspectives, and unsolved questions. *Crit. Rev. Oncol. Hematol.* **2022**, *178*, 103795. [CrossRef]
15. Elahi Najafi, M.A.; Yasui, M.; Teramoto, Y.; Tatenuma, T.; Jiang, G.; Miyamoto, H. GABBR2 as a Downstream Effector of the Androgen Receptor Induces Cisplatin Resistance in Bladder Cancer. *Int. J. Mol. Sci.* **2023**, *24*, 13733. [CrossRef]
16. Sundi, D.; Collier, K.A.; Yang, Y.; Diaz, D.A.; Pohar, K.S.; Singer, E.A.; Gupta, S.; Carson, W.E., 3rd; Clinton, S.K.; Li, Z.; et al. Roles of Androgen Receptor Signaling in Urothelial Carcinoma. *Cancers* **2024**, *16*, 746. [CrossRef] [PubMed]

17. Li, P.; Chen, J.; Miyamoto, H. Androgen Receptor Signaling in Bladder Cancer. *Cancers* **2017**, *9*, 20. [CrossRef]
18. Vukovic, M.; Chamlati, J.M.; Hennenlotter, J.; Todenhofer, T.; Lutfrenk, T.; Jersinovic, S.; Tsaur, I.; Stenzl, A.; Rausch, S. Interleukin-1beta/Interleukin (IL)-1-Receptor-Antagonist (IL1-RA) Axis in Invasive Bladder Cancer-An Exploratory Analysis of Clinical and Tumor Biological Significance. *Int. J. Mol. Sci.* **2024**, *25*, 2447. [CrossRef]
19. Greten, F.R.; Grivennikov, S.I. Inflammation and Cancer: Triggers, Mechanisms, and Consequences. *Immunity* **2019**, *51*, 27–41. [CrossRef]
20. Oxburgh, L. The Extracellular Matrix Environment of Clear Cell Renal Cell Carcinoma. *Cancers* **2022**, *14*, 4072. [CrossRef]
21. Mlynarczyk, G.; Tokarzewicz, A.; Gudowska-Sawczuk, M.; Mroczko, B.; Novak, V.; Novak, A.; Mitura, P.; Romanowicz, L. MMP-14 Exhibits Greater Expression, Content and Activity Compared to MMP-15 in Human Renal Carcinoma. *Int. J. Mol. Sci.* **2024**, *25*, 8107. [CrossRef] [PubMed]
22. Zhao, Y.W.; Ma, W.; Jiang, F.; Xie, Y.; Tang, L. Upregulation of matrix metalloproteinase 14 (MMP14) is associated with poor prognosis in renal clear cell carcinoma-a bioinformatics analysis. *Transl. Androl. Urol.* **2022**, *11*, 1523–1534. [CrossRef]
23. Wang, J.F.; Gong, Y.Q.; He, Y.H.; Ying, W.W.; Li, X.S.; Zhou, X.F.; Zhou, L.Q. High expression of MMP14 is associated with progression and poor short-term prognosis in muscle-invasive bladder cancer. *Eur. Rev. Med. Pharmacol. Sci.* **2020**, *24*, 6605–6615.
24. Fahey, C.C.; Nebhan, C.A.; York, S.; Davis, N.B.; Hurley, P.J.; Gordetsky, J.B.; Schaffer, K.R. Metastatic Penile Squamous Cell Carcinoma Responsive to Enfortumab Vedotin. *Int. J. Mol. Sci.* **2023**, *24*, 16109. [CrossRef]
25. Heath, E.I.; Rosenberg, J.E. The biology and rationale of targeting nectin-4 in urothelial carcinoma. *Nat. Rev. Urol.* **2021**, *18*, 93–103. [CrossRef]
26. Powles, T.; Rosenberg, J.E.; Sonpavde, G.P.; Loriot, Y.; Duran, I.; Lee, J.L.; Matsubara, N.; Vulsteke, C.; Castellano, D.; Wu, C.; et al. Enfortumab Vedotin in Previously Treated Advanced Urothelial Carcinoma. *N. Engl. J. Med.* **2021**, *384*, 1125–1135. [CrossRef] [PubMed]
27. Klumper, N.; Tran, N.K.; Zschabitz, S.; Hahn, O.; Buttner, T.; Roghmann, F.; Bolenz, C.; Zengerling, F.; Schwab, C.; Nagy, D.; et al. NECTIN4 Amplification Is Frequent in Solid Tumors and Predicts Enfortumab Vedotin Response in Metastatic Urothelial Cancer. *J. Clin. Oncol.* **2024**, *42*, 2446–2455. [CrossRef] [PubMed]
28. Yamada, Y.; Beltran, H. The treatment landscape of metastatic prostate cancer. *Cancer Lett.* **2021**, *519*, 20–29. [CrossRef]
29. Dhas, Y.; Biswas, N.; M, R.D.; Jones, L.D.; Ashili, S. Repurposing metabolic regulators: Antidiabetic drugs as anticancer agents. *Mol. Biomed.* **2024**, *5*, 40. [CrossRef]
30. Luconi, M.; Cantini, G.; Crescioli, C. Repurposing glucose-lowering drugs for cancer therapy. *Trends Cancer* **2025**. [CrossRef]
31. Maurya, S.K.; Chaudhri, S.; Kumar, S.; Gupta, S. Repurposing of Metabolic Drugs Metformin and Simvastatin as an Emerging Class of Cancer Therapeutics. *Pharm. Res.* **2025**, *42*, 49–67. [CrossRef] [PubMed]

International Journal of
Molecular Sciences

MDPI

Communication

Tumor-Agnostic Circulating Tumor DNA Testing for Monitoring Muscle-Invasive Bladder Cancer

Raquel Carrasco [1,2,3], Mercedes Ingelmo-Torres [1,2], Ramón Trullas [4], Fiorella L. Roldán [1,2], Leonardo Rodríguez-Carunchio [5], Lourdes Juez [1], Joan Sureda [1], Antonio Alcaraz [1,2], Lourdes Mengual [1,2,3,*,†] and Laura Izquierdo [1,2,†]

1 Laboratori i Servei d'Urologia, Hospital Clinic de Barcelona, 08036 Barcelona, Spain; racarrasco@recerca.clinic.cat (R.C.); ingelmo@recerca.clinic.cat (M.I.-T.); flroldan@clinic.cat (F.L.R.); ljuez@clinic.cat (L.J.); josureda@clinic.cat (J.S.); aalcaraz@clinic.cat (A.A.); lizquier@clinic.cat (L.I.)

2 Genètica i Tumors Urològics, Fundació de Recerca Clinic Barcelona-Institut d'Investigacions Biomèdiques August Pi i Sunyer (FRCB-IDIBAPS), 08036 Barcelona, Spain

3 Departament de Biomedicina, Facultat de Medicina i Ciències de la Salut, Universitat de Barcelona (UB), 08036 Barcelona, Spain

4 Unitat de Neurobiologia, Institut d'Investigacions Biomèdiques de Barcelona (IIBB/CSIC/IDIBAPS), 08036 Barcelona, Spain; ramon.trullas@iibb.csic.es

5 Servei d'Anatomia Patològica, Hospital Clinic de Barcelona, 08036 Barcelona, Spain; lerodrig@clinic.cat

* Correspondence: lmengual@ub.edu; Tel.: +34-93-227-54-00 (ext. 4820)

† Senior authors.

Abstract: Circulating tumor DNA (ctDNA) has recently emerged as a real-time prognostic and predictive biomarker for monitoring cancer patients. Here, we aimed to ascertain whether tumor-agnostic ctDNA testing would be a feasible strategy to monitor disease progression and therapeutic response in muscle-invasive bladder cancer (MIBC) patients after radical cystectomy (RC). Forty-two MIBC patients who underwent RC were prospectively included. Blood samples from these patients were collected at different follow-up time points. Two specific mutations (*TERT* c.1-124C>T and *ATM* c.1236-2A>T) were analyzed in the patients' plasma samples by droplet digital PCR to determine their ctDNA status. During a median follow-up of 21 months, 24% of patients progressed in a median of six months. ctDNA status was identified as a prognostic biomarker of tumor progression before RC and 4 and 12 months later (HR 6.774, HR 3.673, and HR 30.865, respectively; $p < 0.05$). Lastly, dynamic changes in ctDNA status between baseline and four months later were significantly associated with patient outcomes ($p = 0.045$). In conclusion, longitudinal ctDNA analysis using a tumor-agnostic approach is a potential tool for monitoring MIBC patients after RC. The implementation of this testing in a clinical setting could improve disease management and patients' outcomes.

Keywords: bladder cancer; circulating tumor DNA; droplet digital PCR; prognosis

1. Introduction

Muscle-invasive bladder cancer (MIBC) is a heterogeneous and aggressive disease with a poor prognosis. Neoadjuvant chemotherapy (for tumors that are cisplatin-eligible) coupled with radical cystectomy (RC) and lymphadenectomy is the gold standard treatment for localized MIBC [1]. Specific patients might receive adjuvant therapies to reduce tumor progression [1]. Despite these treatments, approximately 50% of MIBC patients will develop local or distant metastases within two years after RC [1,2]. Currently, treatment response monitoring, as well as the detection of metastatic relapse, is performed via imaging techniques [1]. However, imaging background and lesion size could limit the detection of tumor dissemination [3]. Therefore, the identification of prognostic and predictive biomarkers that help detect metastatic relapse earlier and guide treatment would directly impact MIBC patient outcomes.

Circulating tumor DNA (ctDNA) has recently emerged as a real-time prognostic and predictive biomarker for monitoring cancer patients with different tumor types [4–8], including BC [9–12]. Currently, the detection of ctDNA can be based on two approaches: the tumor-informed approach, which requires the initial genomic profiling of the primary tumor, and the tumor-agnostic approach, which is independent of the genomic knowledge of the patient's prior tumor. Previously, we evaluated the potential prognostic utility of a tumor-informed ctDNA approach to monitor MIBC patients after RC [13]. For this purpose, we analyzed four specific mutations, previously identified in tumor tissue [14], in plasma samples from 37 MIBC patients at different follow-up time points after RC. We found that ctDNA status four months after RC is a prognostic biomarker of patients' clinical outcomes. Moreover, we showed that dynamic ctDNA status can detect tumor progression earlier than imaging techniques [13]. Therefore, we demonstrated that ctDNA status could be a useful tool for monitoring MIBC patients after RC.

Here, we aimed to ascertain whether tumor-agnostic ctDNA testing would be a feasible strategy to monitor tumor progression and therapeutic response in MIBC patients after RC. The great advantage of this approach is that it would allow monitoring of MIBC patients, avoiding individual tumor analysis and patient-specific ctDNA assays, thus shortening and simplifying the methodological protocols and promoting the transfer of the approach to clinical practice.

2. Results

2.1. Clinicopathological Features of the Series

The clinicopathological characteristics of the MIBC patients enrolled in this study are shown in Table 1. During a median follow-up of 21 months (range 6–37 months), 10 (24%) patients progressed (seven with LN+); the median time to progression was six months (range 1–11 months). Three of the ten progressive patients received neoadjuvant chemotherapy, and a fourth received adjuvant chemotherapy; the six remaining progressive patients were unfit for chemotherapy. Overall, 50% (5/10) of progressive patients received treatment during tumor progression. During follow-up, eight (19%) patients died, six (75%) due to the BC. The median time of cancer-specific survival (CSS) was 12 months (range 6–21 months).

Table 1. Clinicopathological features of MIBC patients enrolled in this study.

	TOTAL MIBC N (%)	Progressive MIBC N (%)	Non-Progressive MIBC N (%)
	(N = 42)	(N = 10)	(N = 32)
Gender			
Male	39 (93)	9 (90)	30 (94)
Female	3 (7)	1 (10)	2 (6)
Median Age (yr)	67	69	66
Pathological Stage			
<pT2	19 (45)	2 (20)	17 (53)
pT2	11 (26)	2 (20)	9 (28)
pT3	8 (19)	2 (20)	6 (19)
pT4	4 (10)	4 (40)	-
Lymph Nodes (LN)			
LN+	13 (31)	7 (70)	6 (19)
<pT2	2 (5)	1 (10)	1 (3)
pT2	3 (7)	-	3 (9)
pT3	4 (10)	2 (20)	2 (6)
pT4	4 (10)	4 (40)	-
LN-	29 (69)	3 (30)	26 (81)
Neoadjuvant Chemotherapy	15 (38)	3 (30)	12 (38)
<pT2	10 (24)	1 (10)	9 (28)
pT2	2 (5)	1 (10)	1 (3)
pT3	3 (7)	1 (10)	2 (6)
pT4	-	-	-

Table 1. *Cont.*

	TOTAL MIBC N (%)	Progressive MIBC N (%)	Non-Progressive MIBC N (%)
Adjuvant Chemotherapy	5 (12)	1 (10)	4 (13)
<pT2	-	-	-
pT2	2 (5)	-	2 (6)
pT3	2 (5)	-	2 (6)
pT4	1 (2)	1 (10)	-

Abbreviations: LN; lymph nodes; MIBC; muscle-invasive bladder cancer.

2.2. Analysis of TERT and ATM Mutations in cfDNA

The percentage of patients with *TERT* or *ATM* mutations identified in cfDNA samples according to each follow-up time point is depicted in Figure S1. As shown, there was a higher percentage of patients with *TERT* mutations at all follow-up time points. The mean mutant allele fraction (MAF) of the *TERT* mutation was significantly higher in progressive than in non-progressive MIBC patients at baseline and 12 months later (Figure S2A). In addition, the mean MAF of the *TERT* mutation was also significantly higher in those patients who died due to the BC at baseline and 12 months later (Figure S2B). No significant differences were found for the mean MAF of the *ATM* mutation according to tumor progression or death occurrence at any follow-up time point (Table S1).

2.3. Prognostic Value of ctDNA Status

Cox regression analysis evaluated ctDNA status at each follow-up time point and clinicopathological variable. Univariate Cox regression analysis identified positive ctDNA status as a prognostic biomarker of tumor progression before RC and 4 and 12 months later (HR 6.774, HR 3.673, and HR 30.865, respectively; $p < 0.05$). Figure 1 depicts Kaplan–Meier curves showing that ctDNA status can discriminate between two groups of MIBC patients with a significantly different probability of tumor progression before RC and during patient follow-up. Contrarily, ctDNA status was not found to be a prognostic biomarker of CSS at any follow-up time point.

Overall, the subgroup of MIBC patients harboring the *TERT* c.1-124C>T mutation had a worse clinical outcome than those without the mutation (Figure S3). Specifically, MIBC patients harboring this *TERT* mutation before RC and/or four months later were at higher risk of tumor progression than those patients without this mutation at these follow-up time points (HR 5.761, $p = 0.012$). Additionally, MIBC patients harboring the *TERT* c.1-124C>T mutation at both follow-up time points had the highest risk of tumor progression compared with the other MIBC patients (HR 6.405, $p = 0.002$).

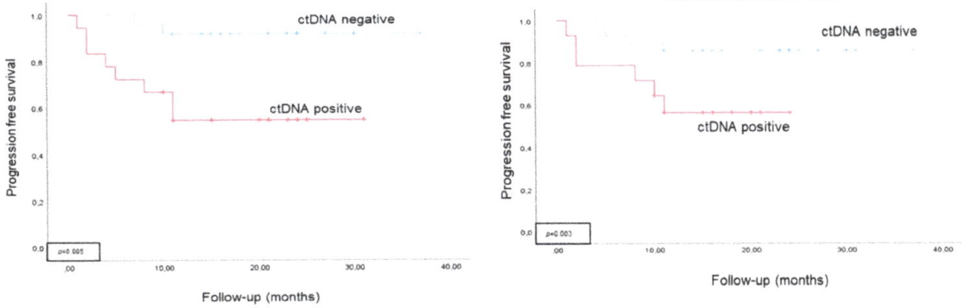

(a) (b)

Figure 1. *Cont.*

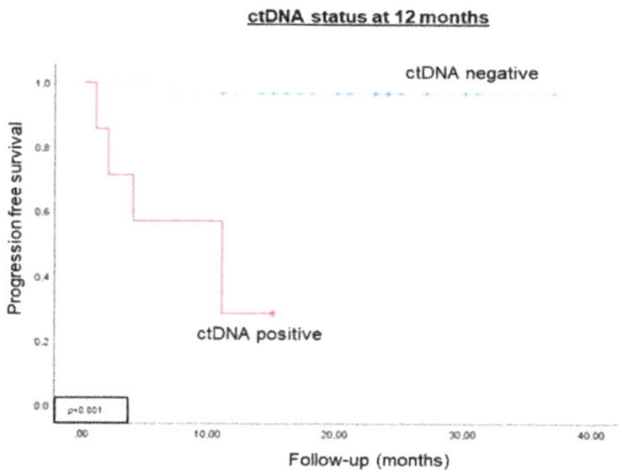

(c)

Figure 1. Prognostic value of ctDNA status at different time points. The Kaplan–Meier survival analysis shows the probability of tumor progression in MIBC patients stratified by ctDNA status (**a**) at baseline, (**b**) four months after RC, and (**c**) 12 months after RC.

2.4. Prognostic Value of Dynamic Changes in ctDNA Status

The patients were stratified into three groups according to their ctDNA status before RC and four months later: Group (1) patients with negative ctDNA status at baseline and at four months; Group (2) patients with positive ctDNA status at baseline and negative ctDNA status four months after RC (ctDNA clearance); and Group (3) patients with positive ctDNA status at four months. As shown in Figure 2, dynamic changes in ctDNA status were significantly associated with the patients' clinical outcomes. In brief, those patients with a positive ctDNA status four months after RC (Group 3) had the poorest clinical outcome. On the other hand, those patients with ctDNA clearance after RC (Group 2) had significantly longer progression-free survival (PFS) than those from Group 3. Finally, patients with undetectable ctDNA in their bloodstream (Group 1) had the best clinical outcome.

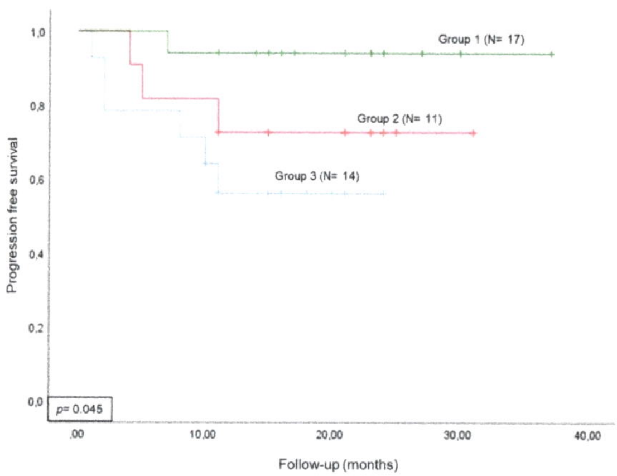

Figure 2. Dynamic ctDNA status analysis. Kaplan–Meier curve for tumor progression in three different MIBC groups stratified by dynamic changes in ctDNA status between baseline and four months after RC: Group 1, patients with negative ctDNA status at baseline and later; Group 2, patients with positive ctDNA status at baseline and negative ctDNA status four months after RC (ctDNA clearance); and Group 3, patients with positive ctDNA status at four months.

3. Discussion

Recently, ctDNA has become a powerful biomarker in oncology because of its utility in monitoring molecular relapse and treatment response or detecting molecular residual disease in multiple cancer types [4,8,15,16], including BC [9,12,17–21]. Using a tumor-informed approach to evaluate serial cfDNA samples from MIBC patients [13], we were able to demonstrate that a positive ctDNA status four months after RC had prognostic implications. We also showed that ctDNA clearance after surgery was significantly associated with patients' clinical outcomes. Moreover, we identified that ctDNA status could predict metastatic relapse earlier than imaging techniques. Therefore, we concluded that ctDNA status could be a potential prognostic biomarker for monitoring MIBC patients after RC. However, this prior study collectively relied on a tumor-informed approach, challenging the translation of these personalized methodological tests to clinical settings because of their evidently high economic cost and workload. The development of a ctDNA assay composed of a panel of the most frequently mutated BC genes in these tumors would avoid assessing patient-specific variants, thus simplifying the methodological process with a single ctDNA assay to monitor all patients.

To the best of our knowledge, the present work is the first to evaluate the prognostic utility of tumor-agnostic ctDNA testing for monitoring MIBC patients after RC using a reduced gene panel targeting a few BC-specific genes. Previous works from other authors on solid tumors such as breast [22], colorectal [23], or lung [24] have already shown that a tumor-agnostic approach with a biomarker panel assay containing the most frequent mutations in these tumors is a feasible strategy for monitoring patients.

Here, in an independent series of MIBC patients, we evaluated the two most frequent somatic mutations found in our previous cohort (69% *TERT* c.1-124C>T and 42% *ATM* c.1236-2A>T) [13]. Our results show that around 40–50% (depending on the follow-up time point) of MIBC patients harbored the *TERT* mutation in their plasma cfDNA samples. Similarly, several studies have described a high percentage of cancer patients, including BC patients, with the *TERT* c.1-124C>T mutation [14,25–29]. Additionally, we found that the mean MAF of *TERT* c.1-124C>T was significantly higher in progressive than non-progressive MIBC patients, as well as in those patients who died due to BC, both before RC and 12 months later, confirming our previous results [13]. These data suggest that radical surgery could reduce the MAF of *TERT* c.1-124C>T in the bloodstream and that an increase in the MAF of this mutation 12 months after RC would indicate the presence of disease relapse. Furthermore, in accordance with other studies analyzing *TERT* mutations in cancer patients [25,30–34], we found that MIBC patients harboring *TERT* c.1-124C>T had a worse clinical outcome than those without the mutation, emphasizing the aggressive role of the *TERT* mutation in MIBC.

The present study also demonstrates that MIBC patients with a positive ctDNA status at baseline and 4 and 12 months later are at high risk of disease progression, confirming our previous ctDNA testing results using a tumor-informed approach [13]. Similarly, Shohdy et al. [19], using a tumor-agnostic 73-gene assay approach (Guardant360® NGS commercial assay), analyzed the ctDNA status of 53 MIBC patients, demonstrating that those patients with a positive ctDNA status at baseline had shorter overall survival.

Finally, we analyzed the dynamic ctDNA status and demonstrated that, after RC, patients with a positive ctDNA status had the worst prognosis, while patients who maintained a negative status had the best clinical outcomes. This change from negative to positive ctDNA status likely indicates a false-negative result at baseline; the use of a larger biomarker panel could settle this discrepancy. On the other hand, ctDNA clearance after surgery was strongly associated with favorable outcomes, according to our previous study and those of other authors [13,19,20,35]. These results further support the usefulness of longitudinal ctDNA analysis to monitor MIBC patients undergoing RC.

The strength of this work lies in the fact that it is the first to demonstrate that a tumor-agnostic approach composed of BC-specific gene assays is a feasible strategy for monitoring MIBC patients after RC. The results obtained herein are comparable with our previous results with a tumor-informed approach. The main advantage of tumor-agnostic ctDNA testing is its easy translation to the clinical setting, as it allows for monitoring of MIBC patients regardless of their patient-specific genomic variants and, therefore, avoids individual tumor analysis and patient-specific ctDNA assays. In addition, the technical approach used in this study, ddPCR, is affordable and usually already present in routine diagnostic laboratories, which would further promote the translation of ctDNA testing to the clinical setting.

We must acknowledge some study limitations. Firstly, we used only two mutations to detect ctDNA in plasma samples; some patients might not harbor these specific mutations. Moreover, the molecular features of MIBC patients can be modified during their follow-up due to, for instance, resistance mechanisms in patients receiving adjuvant treatments. In addition, there may be low levels of ctDNA in plasma at the time of RC due to the non-shedding of the tumor, resulting in a false-negative result at this time point. Finally, although 162 plasma samples were analyzed in this study, our cohort is limited. Therefore, a final validation of our results in a larger independent series is necessary to define the true role of serial ctDNA monitoring in MIBC patients after RC using a tumor-agnostic approach.

4. Materials and Methods

4.1. Patients and Samples

A total of 42 consecutive MIBC patients [median age (range) 67 years (50–80); 39 males, 3 females] who underwent RC and extended lymphadenectomy at our center (Hospital Clinic of Barcelona, Barcelona, Spain) between 2019 and 2022 were prospectively included. The exclusion criterion was the presence of another active neoplasm. Follow-up data were available for all patients. Tumor dissemination was controlled postoperatively via computed tomography scans at three-month intervals for the first year, six-month intervals for the next two years, and annually thereafter. Tumors were considered as progressing when relapses or distant metastases developed during follow-up. PFS and CSS were measured from the date of RC to the event. A final follow-up date was registered in cases without events.

One 10 mL EDTA tube of peripheral blood was collected before RC and 1, 4, and 12 months after surgery. Blood samples were stored at 4 °C until processed within the following 6 h.

All methods were carried out in accordance with relevant guidelines and regulations. All patients provided written informed consent (HCB/2013/8753) before being enrolled in this study. The study methodologies conformed to the standards set by the Declaration of Helsinki and were approved by the Clinical Research Ethics Committee of the Hospital Clinic of Barcelona (HCB/2018/0026).

4.2. Blood Sample Procedures

Blood samples were centrifuged at 3500 rpm for 15 min at 4 °C to separate plasma, followed by plasma centrifugation at 16,000× g for 10 min at 4 °C to remove any remaining cells. The plasma samples were stored at −80 °C until cell-free DNA (cfDNA) extraction.

cfDNA was extracted from 2 to 5 mL of plasma (depending on availability) using the QIAamp Circulating Nucleic Acid kit (Qiagen, Hilden, Germany), according to the manufacturer's instructions, and quantified using the Quant-iT PicoGreen dsDNA kit (Thermo Fisher Scientific, Waltham, MA, USA), in accordance with the manufacturer's instructions.

4.3. Droplet Digital PCR

The determination of ctDNA status in each patient was performed by detecting two specific tumor mutations in their cfDNA samples via droplet digital PCR (ddPCR). These two most frequent somatic mutations identified in our previous study (*TERT* c.1-124C>T and *ATM* c.1236-2A>T) [13] were analyzed in a total of 162 plasma cfDNA samples from 42 MIBC patients before RC and at three time points during the first year of follow-up (1, 4, and 12 months) to determine ctDNA status at each follow-up time point. The ddPCR experiments were performed using the QX200 Droplet Digital OCR system (Bio-Rad Laboratories, Hercules, CA, USA), according to the manufacturer's instructions. The assays used to determine both mutant and wild-type amplicons for each specific mutation were designed and tested as reported in our previous study [13]. Further details about the ddPCR experiments can be checked in our previous study [13].

Data analysis was carried out using QuantaSoft Analysis Pro Software, v1.0 (Bio-Rad Laboratories, Hercules, CA, USA). Original figures pertaining to ddPCR analyzing both *TERT* and *ATM* mutations are shown in Figure S4.

4.4. Statistical Analysis

Differences in the MAF of both *TERT* and *ATM* mutations at different follow-up time points were analyzed using the Mann–Whitney U-test for independent samples and the Friedman test for repeated measures. *TERT* c.1-124C>T and *ATM* c.1236-2A>T mutations were considered positive if the MAF value for each mutation was higher than the cut-off established in our previous study (0.2 for *TERT* and 0.06 for *ATM*) [13]. Additionally, a ctDNA sample was considered positive if at least one of the two mutations had an MAF value higher than these cut-off values. Using this definition, 31% (50/162) of all plasma samples were ctDNA-positive. Kaplan–Meier curves were generated and compared using log-rank tests. A Cox regression analysis was performed on ctDNA status at different follow-up time points and clinicopathological variables (pathological stage, lymph node status, and neo/adjuvant treatments) to examine their influence on PFS and CSS. Additionally, Cox regression analysis was performed on ctDNA status dynamic changes between RC and four months later to examine their influence on PFS. Statistical significance was established at a p-value of 0.05. All analyses were carried out with the SPSS software package (IMB SPSS Statistics 25).

5. Conclusions

Here, we show that serial ctDNA analysis using a tumor-agnostic approach is a powerful tool for monitoring MIBC patients after RC. We found that a positive ctDNA status is a prognostic biomarker of tumor progression at baseline and at 4 and 12 months after RC. Moreover, patients harboring the *TERT* c.1-124C>T mutation before RC and 12 months later are at high risk of recurrence. Finally, we show that dynamic ctDNA status analysis has prognostic implications for MIBC patients after RC. The implementation of tumor-agnostic ctDNA testing in the clinical setting could improve disease management and patients' clinical outcomes.

Supplementary Materials: The following supporting information can be downloaded at: https://www.mdpi.com/article/10.3390/ijms242316578/s1.

Author Contributions: Conceptualization, L.M.; methodology, R.C.; formal analysis, R.C.; investigation, R.C., and L.R.-C.; resources, M.I.-T., R.T., L.J. and J.S.; writing—original draft preparation, R.C.; writing—review and editing, F.L.R., L.M. and L.I.; supervision, A.A., L.M. and L.I.; project administration, L.M.; funding acquisition, L.M. and L.I. All authors have read and agreed to the published version of the manuscript.

Funding: This research was funded by the Instituto de Salud Carlos III (ISCIII) through the Plan Estatal de Investigación Científica y Técnica y de Innovación 2021–2023 (project reference number PI20/00409) and co-funded by the European Regional Development Fund (FEDER).

Institutional Review Board Statement: This study was conducted in accordance with the Declaration of Helsinki and approved by the Clinical Research Ethics Committee of the Hospital Clinic of Barcelona (HCB/2018/0026). All methods were carried out in accordance with relevant guidelines and regulations.

Informed Consent Statement: Informed consent (HCB/2013/8753) was obtained from all subjects involved in this study.

Data Availability Statement: Data are available upon reasonable request.

Acknowledgments: We thank all of the patients who participated in this study and all of the staff and nurses of the Urology Department of the Hospital Clinic for collaborating on sample collection. We thank Helena Kruyer for the English correction of the manuscript. Funding from the CERCA Programme/Generalitat de Catalunya is gratefully acknowledged. This work was developed at the Centre de Recerca Biomèdica Cellex, Barcelona.

Conflicts of Interest: The authors declare no conflict of interest.

References

1. Witjes, J.A.; Bruins, H.M.; Cathomas, R.; Compérat, E.; Cowan, N.C.; Efstathiou, J.A.; Fietkau, R.; Gakis, G.; Hernández, V.; Lorch, A.; et al. European Association of Urology Guidelines on Muscle Invasive and Metastatic Bladder Cancer: Summary of the 2020 Guidelines. *Eur. Urol.* **2021**, *79*, 82–104. [CrossRef]
2. del Pozo Jiménez, G.; Herranz Amo, F.; Arranz Arija, J.A.; Rodríguez Fernández, E.; Subirá Ríos, D.; Lledó García, E.; Bueno Chomón, G.; Cancho Gil, M.J.; Carballido Rodríguez, J.; Hernández Fernández, C. Effect of Adjuvant Chemotherapy in Locally Advanced Urothelial Carcinoma of the Bladder Treated with Cystectomy. *Actas Urol. Esp. (Eng. Ed.)* **2020**, *44*, 94–102. [CrossRef]
3. Paik, M.L.; Scolieri, M.J.; Brown, S.L.; Spirnak, J.P.; Resnick, M.I. Limitations of Computerized Tomography in Staging Invasive Bladder Cancer before Radical Cystectomy. *J. Urol.* **2000**, *163*, 1693–1696. [CrossRef]
4. Zhang, X.; Ju, S.; Wang, X.; Cong, H. Advances in Liquid Biopsy Using Circulating Tumor Cells and Circulating Cell-Free Tumor DNA for Detection and Monitoring of Breast Cancer. *Clin. Exp. Med.* **2019**, *19*, 271–279. [CrossRef]
5. Toledano-Fonseca, M.; Cano, M.T.; Inga, E.; Rodríguez-Alonso, R.; Gómez-España, M.A.; Guil-Luna, S.; Mena-Osuna, R.; de la Haba-Rodríguez, J.R.; Rodríguez-Ariza, A.; Aranda, E. Circulating Cell-Free Dna-Based Liquid Biopsy Markers for the Non-Invasive Prognosis and Monitoring of Metastatic Pancreatic Cancer. *Cancers* **2020**, *12*, 1754. [CrossRef]
6. Vandeputte, C.; Kehagias, P.; El Housni, H.; Ameye, L.; Laes, J.F.; Desmedt, C.; Sotiriou, C.; Deleporte, A.; Puleo, F.; Geboes, K.; et al. Circulating Tumor DNA in Early Response Assessment and Monitoring of Advanced Colorectal Cancer Treated with a Multi-Kinase Inhibitor. *Oncotarget* **2018**, *9*, 17756–17769. [CrossRef]
7. Luo, W.; Rao, M.; Qu, J.; Luo, D. Applications of Liquid Biopsy in Lung Cancer-Diagnosis, Prognosis Prediction, and Disease Monitoring. *Am. J. Transl. Res.* **2018**, *10*, 3911–3923.
8. Yang, Y.; Zhang, T.; Wang, J.; Wang, J.; Xu, Y.; Zhao, X.; Ou, Q.; Shao, Y.; Wang, X.; Wu, Y.; et al. The Clinical Utility of Dynamic ctDNA Monitoring in Inoperable Localized NSCLC Patients. *Mol. Cancer* **2022**, *21*, 117. [CrossRef]
9. Birkenkamp-Demtröder, K.; Christensen, E.; Nordentoft, I.; Knudsen, M.; Taber, A.; Høyer, S.; Lamy, P.; Agerbæk, M.; Jensen, J.B.; Dyrskjøt, L. Monitoring Treatment Response and Metastatic Relapse in Advanced Bladder Cancer by Liquid Biopsy Analysis. *Eur. Urol.* **2018**, *73*, 535–540. [CrossRef]
10. Christensen, E.; Birkenkamp-Demtröder, K.; Sethi, H.; Shchegrova, S.; Salari, R.; Nordentoft, I.; Wu, H.T.; Knudsen, M.; Lamy, P.; Lindskrog, S.V.; et al. Early Detection of Metastatic Relapse and Monitoring of Therapeutic Efficacy by Ultra-Deep Sequencing of Plasma Cell-Free DNA in Patients with Urothelial Bladder Carcinoma. *J. Clin. Oncol.* **2019**, *37*, 1547–1557. [CrossRef]
11. Christensen, E.; Nordentoft, I.; Vang, S.; Birkenkamp-Demtröder, K.; Jensen, J.B.; Agerbæk, M.; Pedersen, J.S.; Dyrskjøt, L. Optimized Targeted Sequencing of Cell-Free Plasma DNA from Bladder Cancer Patients. *Sci. Rep.* **2018**, *8*, 1917. [CrossRef]
12. Pritchard, J.J.G.; Hamilton, G.; Hurst, C.D.; Fraser, S.; Orange, C.; Knowles, M.A.; Jones, R.J.; Leung, H.Y.; Iwata, T. Monitoring of Urothelial Cancer Disease Status after Treatment by Digital Droplet PCR Liquid Biopsy Assays. *Urol. Oncol.* **2020**, *38*, e1–e737. [CrossRef]
13. Carrasco, R.; Ingelmo-Torres, M.; Gómez, A.; Trullas, R.; Roldán, F.L.; Ajami, T.; Moreno, D.; Rodríguez-Carunchio, L.; Alcaraz, A.; Izquierdo, L.; et al. Cell-Free DNA as a Prognostic Biomarker for Monitoring Muscle-Invasive Bladder Cancer. *Int. J. Mol. Sci.* **2022**, *23*, 11732. [CrossRef]
14. Carrasco, R.; Ingelmo-Torres, M.; Gómez, A.; Roldán, F.L.; Segura, N.; Ribal, M.J.; Alcaraz, A.; Izquierdo, L.; Mengual, L. Prognostic Implication of TERT Promoter Mutation and Circulating Tumor Cells in Muscle-Invasive Bladder Cancer. *World J. Urol.* **2022**, *40*, 2033–2039. [CrossRef]
15. Raja, R.; Kuziora, M.; Brohawn, P.Z.; Higgs, B.W.; Gupta, A.; Dennis, P.A.; Ranade, K. Early Reduction in CtDNA Predicts Survival in Patients with Lung and Bladder Cancer Treated with Durvalumab. *Clin. Cancer Res.* **2018**, *24*, 6212–6222. [CrossRef]
16. Magbanua, M.J.M.; Swigart, L.B.; Wu, H.T.; Hirst, G.L.; Yau, C.; Wolf, D.M.; Tin, A.; Salari, R.; Shchegrova, S.; Pawar, H.; et al. Circulating Tumor DNA in Neoadjuvant-Treated Breast Cancer Reflects Response and Survival. *Ann. Oncol.* **2021**, *32*, 229–239. [CrossRef]
17. Tan, M.P.; Attard, G.; Huddart, R.A. Circulating Tumour DNA in Muscle-Invasive Bladder Cancer. *Int. J. Mol. Sci.* **2018**, *19*, 2568. [CrossRef]
18. Powles, T.; Assaf, Z.J.; Davarpanah, N.; Banchereau, R.; Szabados, B.E.; Yuen, K.C.; Grivas, P.; Hussain, M.; Oudard, S.; Gschwend, J.E.; et al. CtDNA Guiding Adjuvant Immunotherapy in Urothelial Carcinoma. *Nature* **2021**, *595*, 432–437. [CrossRef]
19. Shohdy, K.S.; Villamar, D.M.; Cao, Y.; Trieu, J.; Price, K.S.; Nagy, R.; Tagawa, S.T.; Molina, A.M.; Sternberg, C.N.; Nanus, D.M.; et al. Serial CtDNA Analysis Predicts Clinical Progression in Patients with Advanced Urothelial Carcinoma. *Br. J. Cancer* **2022**, *126*, 430–439. [CrossRef]

20. Crupi, E.; de Padua, T.C.; Marandino, L.; Raggi, D.; Dyrskjøt, L.; Spiess, P.E.; Sonpavde, G.P.; Kamat, A.M.; Necchi, A. Circulating Tumor DNA as a Predictive and Prognostic Biomarker in the Perioperative Treatment of Muscle-Invasive Bladder Cancer: A Systematic Review. *Eur. Urol. Oncol.* **2023**. [CrossRef]
21. Nawaf, C.; Shiang, A.; Chauhan, P.S.; Chaudhuri, A.A.; Agarwal, G.; Smith, Z.L. Circulating tumor DNA based minimal residual disease detection and adjuvant treatment decision-making for muscle-invasive bladder cancer guided by modern clinical trials. *Transl. Oncol.* **2023**, *37*, 1936–5233. [CrossRef]
22. Li, S.; Lai, H.; Liu, J.; Liu, Y.; Jin, L.; Li, Y.; Liu, F.; Gong, Y.; Guan, Y.; Yi, X.; et al. Circulating Tumor DNA Predicts the Response and Prognosis in Patients With Early Breast Cancer Receiving Neoadjuvant Chemotherapy. *JCO Precis. Oncol.* **2020**, *4*, 244–257. [CrossRef]
23. Parikh, A.R.; van Seventer, E.E.; Siravegna, G.; Hartwig, A.V.; Jaimovich, A.; He, Y.; Kanter, K.; Fish, M.G.; Fosbenner, K.D.; Miao, B.; et al. Minimal Residual Disease Detection Using a Plasma-Only Circulating Tumor DNA Assay in Patients with Colorectal Cancer. *Clin. Cancer Res.* **2021**, *27*, 5586–5594. [CrossRef]
24. Goldberg, S.B.; Narayan, A.; Kole, A.J.; Decker, R.H.; Teysir, J.; Carriero, N.J.; Lee, A.; Nemati, R.; Nath, S.K.; Mane, S.M.; et al. Early Assessment of Lung Cancer Immunotherapy Response via Circulating Tumor DNA. *Clin. Cancer Res.* **2018**, *24*, 1872–1880. [CrossRef]
25. Günes, C.; Wezel, F.; Southgate, J.; Bolenz, C. Implications of TERT Promoter Mutations and Telomerase Activity in Urothelial Carcinogenesis. *Nat. Rev. Urol.* **2018**, *15*, 386–393. [CrossRef]
26. Leão, R.; Lee, D.; Figueiredo, A.; Hermanns, T.; Wild, P.; Komosa, M.; Lau, I.; Mistry, M.; Nunes, N.M.; Price, A.J.; et al. Combined Genetic and Epigenetic Alterations of the TERT Promoter Affect Clinical and Biological Behavior of Bladder Cancer. *Int. J. Cancer* **2019**, *144*, 1676–1684. [CrossRef]
27. Robertson, A.G.; Kim, J.; Al-Ahmadie, H.; Bellmunt, J.; Guo, G.; Cherniack, A.D.; Hinoue, T.; Laird, P.W.; Hoadley, K.A.; Akbani, R.; et al. Comprehensive Molecular Characterization of Muscle Invasive Bladder Cancer. *Cell* **2017**, *171*, 540–556. [CrossRef]
28. Barthel, F.P.; Wei, W.; Tang, M.; Martinez-Ledesma, E.; Hu, X.; Amin, S.B.; Akdemir, K.C.; Seth, S.; Song, X.; Wang, Q.; et al. Systematic Analysis of Telomere Length and Somatic Alterations in 31 Cancer Types. *Nat. Genet.* **2017**, *49*, 349–357. [CrossRef]
29. Vinagre, J.; Almeida, A.; Pópulo, H.; Batista, R.; Lyra, J.; Pinto, V.; Coelho, R.; Celestino, R.; Prazeres, H.; Lima, L.; et al. Frequency of TERT Promoter Mutations in Human Cancers. *Nat. Commun.* **2013**, *4*, 2185. [CrossRef]
30. Rachakonda, P.S.; Hosen, I.; De Verdier, P.J.; Fallah, M.; Heidenreich, B.; Ryk, C.; Wiklund, N.P.; Steineck, G.; Schadendorf, D.; Hemminki, K.; et al. TERT Promoter Mutations in Bladder Cancer Affect Patient Survival and Disease Recurrence through Modification by a Common Polymorphism. *Proc. Natl. Acad. Sci. USA* **2013**, *110*, 17426–17431. [CrossRef]
31. McEvoy, A.C.; Calapre, L.; Pereira, M.R.; Giardina, T.; Robinson, C.; Khattak, M.A.; Meniawy, T.M.; Pritchard, A.L.; Hayward, N.K.; Amanuel, B.; et al. Sensitive Droplet Digital PCR Method for Detection of TERT Promoter Mutations in Cell Free DNA from Patients with Metastatic Melanoma. *Oncotarget* **2017**, *8*, 78890–78900. [CrossRef]
32. Hirai, M.; Kinugasa, H.; Nouso, K.; Yamamoto, S.; Terasawa, H.; Onishi, Y.; Oyama, A.; Adachi, T.; Wada, N.; Sakata, M.; et al. Prediction of the Prognosis of Advanced Hepatocellular Carcinoma by TERT Promoter Mutations in Circulating Tumor DNA. *J. Gastroenterol. Hepatol.* **2021**, *36*, 1118–1125. [CrossRef]
33. Oversoe, S.K.; Clement, M.S.; Pedersen, M.H.; Weber, B.; Aagaard, N.K.; Villadsen, G.E.; Grønbæk, H.; Hamilton-Dutoit, S.J.; Sorensen, B.S.; Kelsen, J. TERT Promoter Mutated Circulating Tumor DNA as a Biomarker for Prognosis in Hepatocellular Carcinoma. *Scand J. Gastroenterol.* **2020**, *55*, 1433–1440. [CrossRef]
34. Isharwal, S.; Audenet, F.; Drill, E.; Pietzak, E.J.; Iyer, G.; Ostrovnaya, I.; Cha, E.; Donahue, T.; Arcila, M.; Jayakumaran, G.; et al. Prognostic Value of TERT Alterations, Mutational and Copy Number Alterations Burden in Urothelial Carcinoma. *Eur. Urol. Focus* **2019**, *5*, 201–204. [CrossRef]
35. Christensen, E.; Nordentoft, I.; Birkenkamp-Demtröder, K.; Elbæk, S.K.; Lindskrog, S.V.; Taber, A.; Andreasen, T.G.; Strandgaard, T.; Knudsen, M.; Lamy, P.; et al. Cell-Free Urine and Plasma DNA Mutational Analysis Predicts Neoadjuvant Chemotherapy Response and Outcome in Patients with Muscle-Invasive Bladder Cancer. *Clin. Cancer Res.* **2023**, *29*, 1582–1591. [CrossRef]

International Journal of
Molecular Sciences

MDPI

Review

Tumor-Educated Platelets in Urological Tumors: A Novel Biosource in Liquid Biopsy

Mariona Figols [1,2,3], Sviatoslav Chekhun [4,5,6], Maria Fernández-Saorin [4,5], Ignacio Pérez-Criado [1], Ana Bautista [1], Albert Font [4,5,6] and Vicenç Ruiz de Porras [4,5,7,*]

1 Medical Oncology Department, Althaia Xarxa Assistencial Universitària de Manresa, C/ Dr. Joan Soler, 1-3, 08243 Manresa, Spain; mfigols@althaia.cat (M.F.); iperezc@althaia.cat (I.P.-C.); ambautista@althaia.cat (A.B.)
2 PhD Programme in Medicine and Biomedical Sciences, Doctoral School, University of Vic, Central University of Catalonia (UVic-UCC), C/ Dr. Junyent, 1, 08500 Vic, Spain
3 Faculty of Medicine, University of Vic, Central University of Catalonia (UVicUCC), Can Baumann, Ctra, de Roda, 70, 08500 Vic, Spain
4 CARE Program, Germans Trias i Pujol Research Institute (IGTP), Camí de les Escoles, s/n, 08916 Badalona, Spain; schekhun.germanstrias@gencat.cat (S.C.); mfernandezs@igtp.cat (M.F.-S.); afont@iconcologia.net (A.F.)
5 Badalona Applied Research Group in Oncology (B·ARGO), Catalan Institute of Oncology, Camí de les Escoles, s/n, 08916 Badalona, Spain
6 Medical Oncology Department, Catalan Institute of Oncology, Camí de les Escoles, s/n, 08916 Badalona, Spain
7 GRET and Toxicology Unit, Department of Pharmacology, Toxicology and Therapeutic Chemistry, Faculty of Pharmacy and Food Sciences, University of Barcelona, 08028 Barcelona, Spain
* Correspondence: vruiz@igtp.cat; Tel.:+34-606587010

Abstract: Platelets, traditionally recognized for their role in hemostasis, have emerged as pivotal players in cancer biology. They actively contribute to tumor proliferation, angiogenesis, immune evasion, and metastasis and thus play a significant role in cancer progression. Tumor-educated platelets (TEPs) acquire protumorigenic phenotypes through RNA, protein, and receptor profile alterations driven by interactions with tumors and their microenvironment. These modifications enable TEPs to enhance tumor growth and dissemination and to play a critical role throughout the metastatic process. Moreover, TEPs are promising biomarkers that can easily be analyzed in liquid biopsies. Since they dynamically mirror tumor activity through transcriptomic and proteomic changes, their analysis offers a non-invasive method for determining cancer detection and diagnosis, patient prognosis, therapy monitoring, and personalization of treatment. Their demonstrated accuracy in identifying cancer types and predicting treatment responses underscores their ability to provide real-time insights into tumor biology, including in urological malignancies. Their diagnostic potential and their accessibility as blood-sourced biomarkers position TEPs as transformative tools in advancing personalized oncology. Here, we focus on the role of TEPs in urological tumors, exploring their applications in early cancer detection, disease monitoring, and the design of tailored therapeutic strategies.

Keywords: tumor-educated platelets; liquid biopsy; biomarkers; urological tumors

1. The Role of Platelets in Cancer: An Overview

Cancer is a leading cause of mortality worldwide, with its progression involving complex interactions between tumor cells and various components of the host's circulatory and immune systems. Among these components, platelets have emerged as critical players in cancer biology [1].

Platelets, the smallest anucleated elements of the blood system, originate from megakaryocytes. While their role has traditionally been observed mainly in hemostasis and coagulation, they are now recognized as dynamic participants in various physiological and pathological processes, including inflammation and immune response [2]. They possess a rich array of surface receptors, and when activated, they release numerous bioactive molecules stored in their alpha-granules and dense granules. These molecules include growth factors, cytokines, chemokines, and adhesion molecules, which can modulate the behavior of other cells, including tumor cells. The interaction between tumors and platelets significantly influences various aspects of cancer development, including tumor proliferation, angiogenesis, immune evasion, and metastasis [3,4] (Figure 1). The subsequent sections present a broad review of the multiple facets of platelets in cancer development.

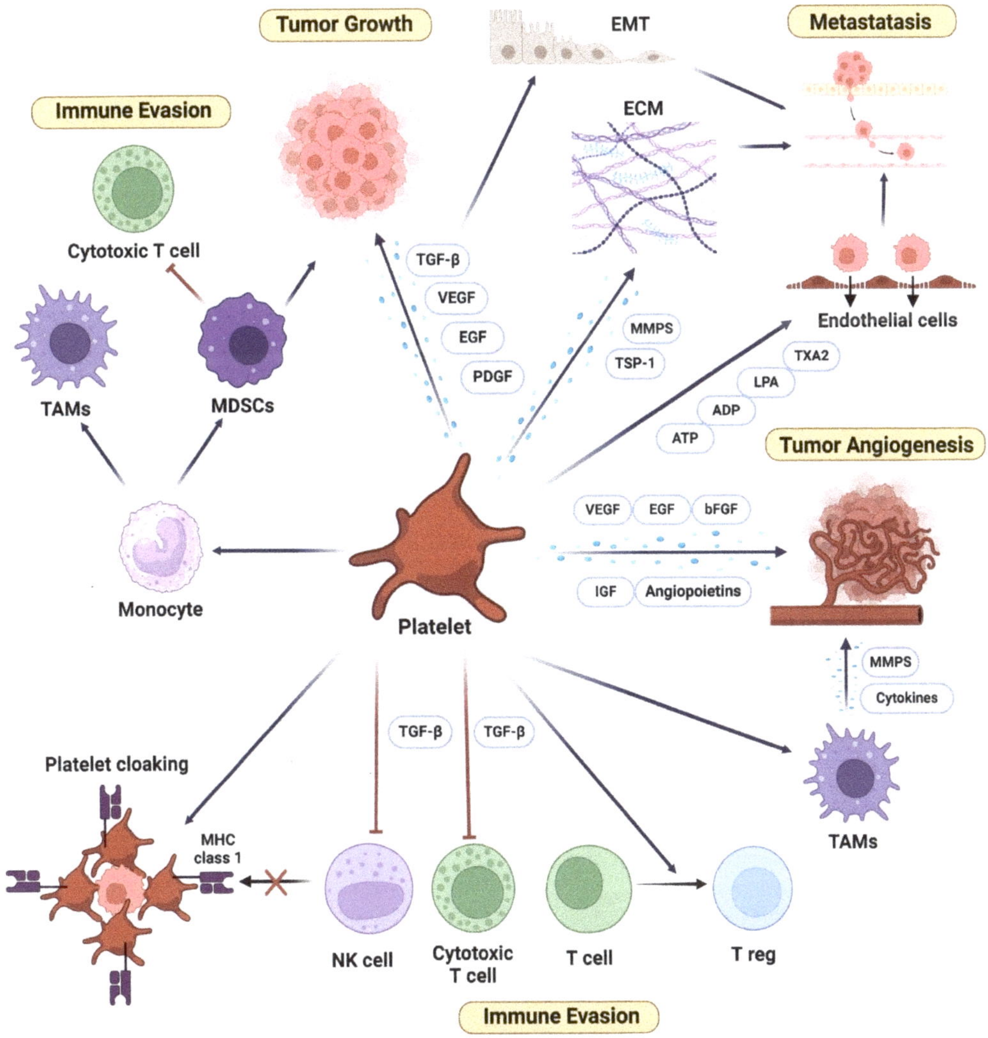

Figure 1. Platelets significantly influence various aspects of cancer development, including tumor proliferation, angiogenesis, immune evasion, and metastasis. These processes are mediated through

direct or indirect interactions with tumor cells and the tumor microenvironment, involving the secretion of growth factors, chemokines, and the modulation of immune cells, which play pivotal roles in shaping the tumor microenvironment and driving cancer progression. bFGF: Basic Fibroblast Growth Factor; ECM: Extracellular Matrix; EGF: Epidermal Growth Factor; EMT: Epithelial–Mesenchymal Transition; IGF: Insulin Growth Factor; LMA: Lysophosphatidic Acid MDSCs: Myeloid-Derived Suppressor Cells; MHC: Major Histocompatibility Complex; MMPs: Metalloproteinases; NK Cell: Natural Killer Cell; PDGF: Platelet-Derived Growth Factor; TAMs: Tumor-Associated Macrophages; TGF-β: Transforming Growth Factor-Beta; Treg: Regulatory T Cell; TSP-1: Thrombospondin-1; TXA2: Thromboxane A2; VEGF: Vascular Endothelial Growth Factor. ⊢: Inhibition; ←: Activation. Created in https://BioRender.com (accessed on 7 January 2025).

1.1. Platelets in Tumor Growth and Progression

Platelets are actively involved in promoting tumor growth and progression through various mechanisms. One of the key ways that they facilitate tumor growth is by secreting a wide range of growth factors, cytokines, and chemokines, which create a favorable microenvironment for tumor cells [5,6].

Platelet–tumor cell interactions can occur directly, facilitated by various adhesion molecules such as P-selectin, glycoprotein IIb/IIIa (GPIIb/IIIa), integrins, and CD40 ligand (CD40L). These interactions are crucial for the survival and dissemination of tumor cells. For example, P-selectin on platelets binds to its ligand on tumor cells, facilitating tumor cell arrest on the vascular endothelium, which is a critical step in tumor progression and metastasis [7].

Enhanced platelet activation is evident in patients with bladder cancer, as demonstrated by increased surface P-selectin expression and elevated levels of its soluble form. This activation plays a significant role in tumor growth and progression. A research team from Cairo University and the Theodor Bilharz Research Institute in Egypt evaluated 30 bladder cancer patients using flow cytometry and ELISA to measure these markers. Their findings suggest that P-selectin-mediated interactions contribute critically to tumor progression, highlighting this axis as a promising target for therapeutic intervention [8]

Platelets store and release a variety of growth factors, including platelet-derived growth factor (PDGF), transforming growth factor-beta (TGF-β), vascular endothelial growth factor (VEGF), and epidermal growth factor (EGF), all of which play crucial roles in cell proliferation, survival, and migration. In particular, PDGF and TGF-β enhance tumor cell proliferation and survival by activating signaling pathways that promote cell cycle progression and inhibit apoptosis [9]. The release of these growth factors by platelets can be triggered by various stimuli, including direct interaction with tumor cells, exposure to tumor-derived microparticles, and activation by thrombin or collagen [10].

1.2. Platelets in Tumor Angiogenesis

Angiogenesis, the growth of new blood vessels from existing structures, is a key factor in cancer, as these new vessels supply oxygen and nutrients that promote tumor growth and metastasis [11]. Platelets play a crucial role in angiogenesis by serving as reservoirs and active mediators of pro-angiogenic factors, thereby influencing the tumor microenvironment (TME). Platelets are rich in angiogenic molecules, such as VEGF, EGF, basic fibroblast growth factor (bFGF), insulin growth factors, and angiopoietins. Upon activation, platelets release these factors, which promote endothelial cell proliferation, migration, and tube formation—fundamental processes in the formation of new blood vessels. This dynamic contribution underscores the central role of platelets in driving angiogenesis, particularly in the context of tumor progression [12–14].

Renal cell carcinoma (RCC) is a paradigmatic tumor in its reliance on angiogenesis. In this context, platelets perform specialized functions that extend beyond their conventional pro-angiogenic role. They not only deliver VEGF to stimulate endothelial proliferation and tube formation but also transport PDGF-B, which acts synergistically with VEGF to stabilize the newly formed vessels [15]. Through the PDGF-B/PDGFRβ axis, platelets recruit pericytes and smooth muscle cells, promoting rapid vascular maturation [16]. This phenomenon is particularly notable in clear cell carcinomas and is associated with poorer clinical outcomes [17].

In addition to releasing pro-angiogenic factors, platelets can also modulate the TME in other ways to promote angiogenesis. They interact with endothelial cells, pericytes, fibroblasts, and immune cells in the TME [13,14], and they release cytokines and chemokines, creating favorable conditions for angiogenesis and tumor progression [13,14]. In addition, platelets can recruit and activate tumor-associated macrophages (TAMs), which also promote angiogenesis and tumor progression. TAMs release additional pro-angiogenic factors, such as matrix metalloproteinases (MMPs) and cytokines, further enhancing angiogenesis [18,19].

1.3. *Platelets and Immune Evasion in Cancer*

One of the key challenges in cancer therapy is the ability of tumor cells to evade the host immune system. Cancer cells have evolved numerous strategies to avoid immune detection and destruction, and platelets play a crucial role in protecting tumor cells from immune system attacks. Though not classified as immune system cells, platelets are extensively involved in the functions of several immune cells, including natural killer (NK) cells, monocytes, and certain T lymphocytes. Furthermore, the cytokines and other mediators released by platelets can suppress immune responses [20,21]. In addition, platelets can form aggregates with tumor cells, protecting them from shear stress and immune surveillance in the bloodstream [21,22]. This phenomenon, known as "platelet cloaking", shields circulating tumor cells (CTCs) from NK cell-mediated cytotoxicity, thereby enhancing their survival and metastatic potential. This cloaking is mediated by the expression of "self" antigens on platelets, such as major histocompatibility complex (MHC) class I molecules, which mask the tumor cells and prevent their recognition by NK cells [21,23], a process that is crucial for protection against the immune system [24]. Moreover, platelets promote the expression of immune checkpoint molecules such as programmed death ligand 1 (PD-L1) on tumor cells, which also facilitates immune evasion [21,25].

Beyond platelet cloaking, platelets actively suppress T-cell responses and undermine T-cell-based immunotherapies. They impair these treatments—such as bispecific antibodies targeting PSMA—by reducing the recruitment and effector functions of CD4+ and CD8+ T cells. Platelet activation through TGF-β signaling leads to diminished T-cell degranulation, perforin release, and target cell lysis, ultimately compromising therapeutic efficacy. Targeting platelet function or inhibiting the TGF-β pathway may help restore T-cell reactivity and enhance treatment outcomes [26].

Platelets play a multifaceted role in modulating immune cell behavior within the TME, contributing to an immunosuppressive state that supports tumor progression. They release immunosuppressive cytokines like TGF-β, which inhibits the activation and function of cytotoxic T cells and NK cells, thereby impairing the immune response against tumor cells. Additionally, platelet-derived TGF-β can convert effector T cells into regulatory T cells (Tregs), which further suppress the immune response and promote tumor tolerance [21,27]. In this context, the TGFβ-docking receptor glycoprotein A repetitions-predominant (GARP)–TGF-β axis plays a pivotal role in tumor immune evasion by activating latent TGF-β on platelets and Tregs, thus promoting immune suppression and tumor progression [28–30]. Platelets also influence monocyte differentiation, inducing the formation of myeloid-derived suppressor cells (MDSCs), which inhibit T-cell responses and drive tumor growth [31].

In bladder cancer, platelets interact with stromal components of the TME, such as macrophages and fibroblasts, to promote an immunosuppressive microenvironment. For instance, they enhance the recruitment of M2 macrophages, which suppress antitumor immunity and contribute to tumor aggressiveness [32].

Moreover, platelets can promote the differentiation of monocytes into TAMs with a pro-tumorigenic M2 phenotype, enhancing the immune evasion capability of the tumor [21]. In summary, by shaping immune cell behavior, platelets play a critical role in facilitating immune evasion and supporting tumor growth and metastasis.

1.4. Platelets in Cancer Metastasis

Metastasis is the process by which cancer cells spread from the primary tumor to other sites in the body. This complex, multistep process involves the detachment of tumor cells from the primary site, invasion into surrounding tissues, intravasation into the bloodstream, survival within the circulatory system, extravasation at a distant location, and colonization of a new organ [33]. When tumor cells enter the bloodstream, they encounter a hostile environment characterized by shear stress and immune surveillance. Platelets play a protective role through platelet cloaking, which not only shields the tumor cells from these hostile forces but also enhances the adhesive properties of CTCs, thus facilitating their interaction with the vascular endothelium and promoting extravasation into distant tissues [9,22,23,34].

This extravasation is a critical step in metastasis [33], and platelets play a pivotal role in facilitating this process by interacting with endothelial cells and promoting the formation of gaps in the endothelial barrier, thereby enabling tumor cells to exit the bloodstream [35]. To achieve this vascular permeability, platelets release factors such as ADP, thromboxane A2, and ATP, which activate endothelial cells, thus supporting the metastatic dissemination of tumor cells [7]. Moreover, platelets secrete lysophosphatidic acid, which enhances tumor cell adhesion to endothelial cells and promotes their transendothelial migration [36].

For example, in prostate cancer, platelet-derived lysophosphatidic acid (LPA) and CCL3L1 enhance tumor cell adhesion and survival via receptors such as LPAR3 and CCR1, which are notably overexpressed in aggressive disease. Interactions between platelets and prostate cancer cells through ephrin–EPH receptor signaling promote apoptotic resistance, while platelet-derived microvesicles transfer pro-invasive factors to tumor cells, further amplifying their metastatic potential. These mechanisms illustrate how platelets actively contribute to the metastatic cascade by recruiting pro-metastatic cells and facilitating bidirectional signaling that enhances tumor cell survival and invasion [37,38].

Additionally, platelets release matrix MMPs and thrombospondin-1 (TSP-1), which play critical roles in remodeling the extracellular matrix (ECM) and degrading the basement membrane. MMPs, including MMP-2 and MMP-9, are proteolytic enzymes that cleave key ECM proteins like collagen and fibronectin, creating pathways for tumor cells to invade surrounding tissues and extravasate into distant sites [39]. TSP-1, a multifunctional glycoprotein, not only activates MMP-2 but also interacts with integrins and other ECM components to facilitate cell migration [40,41].

Once tumor cells have extravasated, they need to survive and establish a metastatic niche to colonize the new organ. Platelets contribute to this step in the metastatic process by secreting factors that promote tumor cell survival and dormancy. For example, platelets release TGF-β and other cytokines that inhibit apoptosis and induce a quiescent state in tumor cells, allowing them to evade immune detection and survive in the new microenvironment. Moreover, TGF-β signaling promotes epithelial–mesenchymal transition (EMT) in tumor cells, facilitating their migration and invasion [7]. Furthermore, platelets can create a pre-metastatic niche by interacting with stromal cells and immune cells at distant sites, creating a favorable environment for tumor cell colonization. This includes recruiting bone marrow-derived cells, such as MDSCs and TAMs, which promote tumor cell survival and growth [18].

Taken together, through their multifaceted roles at different stages of the metastatic process—protecting tumor cells, promoting endothelial permeability, and shaping the metastatic niche—platelets serve as critical enablers of metastasis, enhancing the survival, dissemination, and colonization of cancer cells.

2. Tumor-Educated Platelets

In addition to the crucial role of platelets in several stages of cancer progression, tumor-educated platelets (TEPs) have shown great potential as biomarkers for the early detection of cancer, disease monitoring, and development of tailored therapeutic strategies, particularly in solid tumors. Here, we review the emerging potential of TEPs with a specific emphasis on urological malignancies.

TEPs constitute a distinct subpopulation of platelets that have been exposed to tumors. This exposure modifies the RNA and protein profiles of the platelets to reflect the molecular characteristics of the tumor [42]. This process, known as "tumor education", enables platelets to support tumor growth, metastasis, and immune evasion more effectively [43]. This transformation is driven by direct interactions between tumor cells and platelets through adhesion molecules such as P-selectin, integrins, and glycoproteins. Additionally, it involves the activation of platelet receptors by tumor-released extracellular molecules, including thrombin, tissue factors, and extracellular vesicles (EVs). These EVs transfer tumor-derived proteins, lipids, and RNA to the platelets, enabling them to sequester spliced or unspliced mRNA and preserve transcripts from megakaryocytes, thereby altering the platelet transcriptome and behavior. TEPs exhibit significant transcriptomic changes, including the upregulation of PDGF and TGF-β1, which promote angiogenesis and tumor progression. Certain cancer-associated cytokines, such as granulocyte colony stimulating factor (G-CSF), granulocyte macrophage-colony stimulating factor (GM-CSF), and interleukin-6 (IL-6), enhance megakaryopoiesis and increase platelet counts, inducing tumor-educated megakaryocytes that play a pivotal role in the formation of TEPs, particularly under the influence of IL-6. With these adaptations, platelets acquire a protumorigenic phenotype, which enhances their ability to support angiogenesis, tumor invasion, and TME remodeling [44,45] (Figure 2).

In addition to their enhanced role as active mediators of tumor progression, TEPs have shown promise as valuable biomarkers detectable by minimally invasive liquid biopsies. Here, we focus on the analysis of TEPs as a promising tool for cancer detection, monitoring, prognosis, and personalization of treatment [46,47].

2.1. TEPs as Blood-Sourced Biomarkers for Solid Tumors

Platelets cannot synthesize RNA, and instead, their RNA is either endocytosed from blood circulation or derived from megakaryocytes. Tumor-derived signals can induce specific changes in platelets, and the RNA content of TEPs is dynamically regulated by the TME. This "tumor education" allows platelets to internalize tumor-derived biomolecules, including RNA, which confers an RNA profile in the platelets that reflects that of the tumor [44,48] (Figure 2). Interestingly, these changes can be detected and analyzed using high-throughput sequencing techniques [49]. This ability of TEPs to carry tumor-specific information positions them as a promising biomarker that can easily be analyzed in liquid biopsies of patients with solid tumors. Unlike traditional biopsies, which require invasive tissue sampling, liquid biopsies offer a less invasive alternative by utilizing easily accessible body fluids like blood. Moreover, the advantages of TEPs compared to other blood-based biosources—such as CTCs, exosomes, circulating tumor DNA (ctDNA), and cell-free nucleic acids—include their abundance, ease of isolation, stability and capacity to process RNA [50,51]. The analysis of TEPs in liquid biopsies thus provides a non-invasive method for the detection of cancer, monitoring of treatment response, and identification of potential therapeutic targets [42,46–48].

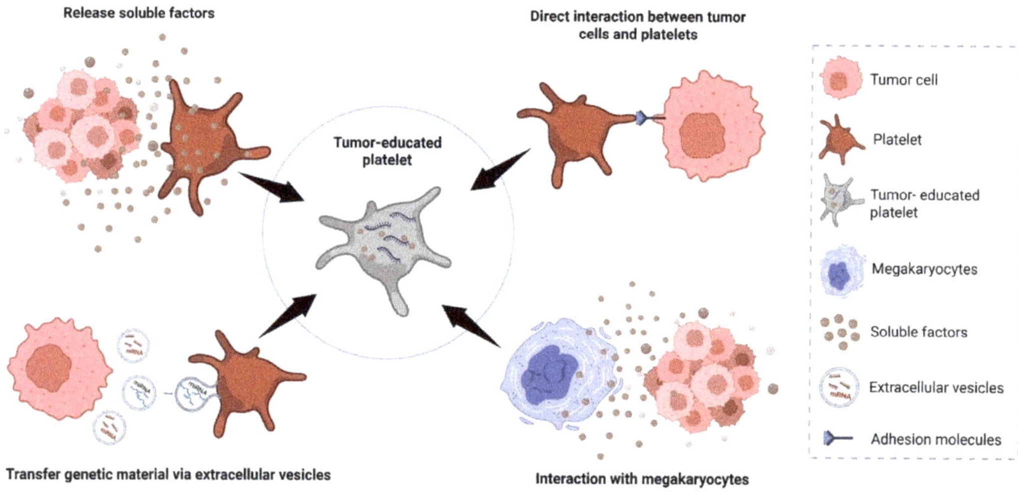

Figure 2. Mechanisms of "tumor education" of platelets. These include the secretion of soluble tumor-derived factors, direct interactions between platelets and tumor cells, the transfer of genetic material via extracellular vesicles, and the interaction with megakaryocytes. Created in https://BioRender.com (accessed on 10 April 2025).

In 2010, Calverley and colleagues were the first to analyze platelet RNA in patients with solid tumors. Using microarray analysis, they profiled platelet mRNA from seven healthy individuals and five patients with untreated metastatic non-small cell lung cancer (NSCLC). Their study identified 200 altered RNAs between the two groups, with 197 transcripts showing decreased levels in the platelets of NSCLC patients. This pilot study provided the first evidence that platelet RNA could serve as a potential diagnostic tool [52].

Nilsson and colleagues later observed that tumor-derived RNA can be transferred into blood platelets and that platelets from prostate and glioma cancer patients contain cancer-associated RNA biomarkers [53]. In 2016, the same group employed RT-PCR to analyze echinoderm microtubule associated protein-like 4 (EML4)-anaplastic lymphoma kinase (ALK) rearrangements in platelets and plasma from 77 NSCLC patients. They observed a 65% sensitivity and 100% specificity for detection. Moreover, in a subgroup of patients with EML4-ALK-rearranged tumors undergoing targeted therapy, the authors tracked changes in EML4-ALK-rearranged RNA levels in platelets over the course of treatment and found that these levels correlated with progression-free survival (PFS) and overall survival (OS). Notably, monitoring platelet EML4-ALK expression predicted resistance to targeted therapy approximately two months before conventional imaging methods detected disease progression [54].

In a salient study by Best and colleagues, mRNA sequencing of TEPs from 283 samples was able to distinguish 228 patients with cancer from 55 healthy individuals with 96% accuracy. Moreover, among six different tumor types (NSCLC, colorectal cancer, glioblastoma, pancreatic cancer, hepatobiliary cancer, and breast cancer), the location of the primary tumor was correctly identified with 71% precision. Furthermore, the study identified several genetic alterations, including MET and ERBB2 positivity and mutations in Kirsten rat sarcoma virus (KRAS), epidermal growth factor receptor (EGFR), phosphatidylinositol-4,5-bisphosphate 3-kinase, and catalytic subunit alpha (PIK3CA) [42].

Along the same lines, Sol et al. demonstrated that TEP-derived RNA signatures, identified using swarm intelligence, were able to detect and monitor glioblastoma. The RNA profiles of TEPs from glioblastoma patients were distinguishable from those of patients with metastatic brain cancer, multiple sclerosis, and healthy controls, with an accuracy exceeding 80% in all cases. Additionally, the study provided evidence that tumor signals in TEPs are dynamic, which enabled pseudoprogression to be differentiated from true progression [55].

Building on the foundational work of these studies, which pioneered the exploration of TEPs and demonstrated their potential as diagnostic, prognostic, monitoring, and therapeutic targeting tools, subsequent research has further emphasized the utility of TEPs across various tumor types, including sarcoma, nasopharyngeal carcinoma, colorectal cancer, and NSCLC. Heinhuis and colleagues focused on the analysis of TEPs in sarcoma. Using RNA-seq, they analyzed TEP samples from patients with various sarcoma subtypes as well as from control groups of former sarcoma patients in remission and from healthy donors. They found significant transcriptional changes in TEPs that were able to distinguish between affected patients and controls with an accuracy exceeding 85%. Additionally, their gene expression analysis suggested that TEPs can differentiate between different sarcoma subtypes. These findings highlight the potential of analyzing TEP RNA in liquid biopsy as a diagnostic tool for sarcoma [56]. In a study of nasopharyngeal carcinoma, Wang et al. identified two TEP microRNAs (miR-34c-3p and miR-18a-5p) that were upregulated in patients compared to healthy donors, suggesting their promise as biomarkers for disease detection. Notably, these expression changes were absent in plasma, highlighting a "tumor education" effect occurring specifically within the platelets [57]. In a study of colorectal cancer, Yang et al. found that TIMP metallopeptidase inhibitor 1 (*TIMP1*) mRNA levels were elevated in platelets from patients compared to healthy individuals or those with inflammatory bowel diseases, indicating a potential as a diagnostic biomarker for colorectal cancer [58]. Most recently, in advanced NSCLC, Hu and colleagues evaluated PD-L1 expression levels in tumor tissue and TEPs in patients treated with immunotherapy. Although expression levels did not correlate between tumor tissue and TEPs, elevated PD-L1 mRNA levels in TEPs were associated with improved outcomes, underscoring their potential as a surrogate biomarker to guide treatment decisions [59].

Taken together, the findings from these studies strongly support the potential of TEPs as a blood-sourced biomarker for detecting and monitoring various tumors. TEPs offer unique advantages, such as their dynamic RNA content, which reflects real-time tumor activity and enables the identification of cancer-specific signatures. In the following sections, we focus on the potential role of TEPs as prognostic and predictive biomarkers in prostate, kidney, and bladder cancers. Understanding the specific alterations in TEP RNA profiles within these urological malignancies could pave the way for improved patient stratification, non-invasive disease monitoring, and enhanced therapeutic decision-making.

Table 1 provides a summary of the main studies conducted on TEPs in prostate, kidney, and bladder cancers. It highlights key findings, methodologies, and clinical implications and offers a comprehensive overview of the current state of research in these urological malignancies.

Table 1. Summary of key studies investigating tumor-educated platelets (TEPs) in prostate, kidney, and bladder cancers.

Study Reference	Objective of TEP Analysis	Gene Expression Signature	Method of Analysis	N° Cases/Controls	Key Findings	Clinical Implications
			Prostate Cancer			
[53]	Detection of tumoral RNA in platelets	*PCA3*	RT-PCR	12 patients/ 10 controls	PCA3 RNA detected in prostate cancer patients	Potential for prostate cancer detection
[60]	Early tumor detection	*PCA3, MALAT1, EZH2, AMACR PSGR, PSA, PSMA, TRPM8*	*RT-PCR*	31 patients/ 29 controls	No significant differences between cases and controls	TEPs may not be suitable for early tumor detection
[61]	Tumor detection and tumor origin identification	Broad pan-cancer RNA	ThromboSeq (RNA-seq)	35 patients	Detected tumor in 92% (AUC = 0.98), improved with higher stage; origin identified in >80%	High potential for detecting advanced prostate cancer
[62]	Prediction of response to therapy	*KLK2, KLK3, FOLH1, NPY*	Digital PCR	50 patients/ 15 controls	A three-gene panel (KLK3, NPY, FOLH1) identified resistance to abiraterone therapy	Predictive/ prognostic tool for mCRPC patients
			Kidney Cancer			
Study Reference	**Objective of TEP Analysis**	**Gene Expression Signature**	**Method of Analysis**	**N° Cases/Controls**	**Key Findings**	**Clinical Implications**
[63]	Diagnostic	68-gene panel	RNA-seq	24 patients/ 25 controls	Diagnostic accuracy of 95.9% (AUC: 0.988)	Significant potential for RCC blood-based screening
[61]	Tumor detection and tumor origin identification	Broad pan-cancer RNA	ThromboSeq (RNA-seq)	28 patients	Detection accuracy of 66% (AUC = 0.87); tumor origin identified in >80%	Demonstrated potential for liquid biopsy though limited renal-specific data
			Bladder Cancer			
Study Reference	**Objective of TEP Analysis**	**Gene Expression Signature**	**Method of Analysis**	**N° Cases/Controls**	**Key Findings**	**Clinical Implications**
[61]	Tumor detection and tumor origin identification	Broad pan-cancer RNA	ThromboSeq (RNA-seq)	28 patients	Detection accuracy of 89% (AUC 0.99); tumor origin identified in >80%	Demonstrated potential for liquid biopsy though limited bladder-specific data
[64]	Predict pathological response to neoadjuvant chemotherapy	Not specified	WES on tissue samples, followed by qPCR on liquid biopsy	150 patients	No results	Potential for bladder preservation by avoiding unnecessary cystectomies

AUC: area under the curve; **mCRPC**: metastatic castration-resistant prostate cancer; **RCC**: renal cell carcinoma; **TEPs**: tumor-educated platelets.

2.2. TEPs and Prostate Cancer

Prostate cancer is one of the most common malignancies in men, second only to lung cancer [65]. Most cases are detected at a localized stage, requiring transrectal or transperineal biopsy, which are invasive procedures with potential complications. In advanced cases, metastases frequently spread to the bones and retroperitoneal lymph nodes, making it difficult to obtain biopsy samples [66]. In fact, bone biopsies are particularly challenging, as the decalcification process required for sample preparation can degrade nucleic acids, limiting the ability to perform molecular analyses [67]. Additionally, tissue biopsies often fail to reflect tumor heterogeneity. Given these limitations, liquid biopsy has emerged as a minimally invasive and highly promising alternative for molecular profiling in prostate cancer [68]. Among the various biosources available, TEPs stand out for their ability to dynamically reflect the molecular characteristics of the tumor.

The first findings on TEPs in prostate cancer were reported in 2011 by Nilsson and colleagues. They identified the presence of EGFR variant III (EGFRvIII) and prostate cancer antigen 3 (PCA3) in platelets isolated from glioma and prostate cancer patients, respectively. Moreover, significant differences in RNA profiles were observed in platelets from glioma patients compared to those of healthy controls, suggesting a distinct cancer signature [53].

Building on Nilsson's findings, Hänze and colleagues conducted a pilot study to evaluate whether TEPs could be used for early detection of prostate cancer. Using relative quantitative RT-PCR, they analyzed RNA in platelets from 31 patients with localized prostate cancer undergoing radical prostatectomy and 29 healthy young men as controls. The study included a variety of RNA markers, such as PCA3 and its isoforms, metastasis-associated lung adenocarcinoma transcript 1 (MALAT1), enhancer of zeste homolog 2 (EZH2), prostate-specific antigen (PSA), and prostate-specific membrane antigen (PSMA). The results revealed heterogeneous expression patterns of RNA markers in platelets, with no statistically significant differences between the groups. PSA was detected in only two patients (10%) and was absent in all control samples, but this difference was not statistically significant. The authors concluded that while TEPs were not effective for early detection in this study, certain RNA markers in platelets, such as PCA3 and MALAT1, may still hold potential as biomarkers for prostate cancer. They emphasized the need for larger, prospective studies to further investigate the role of platelets in prostate cancer diagnosis and to determine whether TEPs can ultimately serve as reliable biomarkers [60].

In this context, In 't Veld and colleagues developed and validated a pan-cancer detection test using platelet RNA analysis, introducing a promising method for cancer detection and localization. Their study utilized a dataset of 2351 samples, which included 18 tumor types, asymptomatic controls, and symptomatic controls without a cancer diagnosis. Their pan-cancer ThromboSeq algorithm was able to distinguish between 18 tumor types with high specificity. Notably, prostate cancer was the most frequently detected tumor type, identified in 92% of cases, with an impressive area under the curve (AUC) of 0.98. Moreover, the accuracy of detection varied with tumor stage, showing improved performance in advanced stages, ranging from 46% for stage I to 72% for stage IV. Furthermore, the study accurately identified the tumor origin in over 80% of cancer patients for five specific tumor types, including a combined category encompassing urothelial, prostate, and renal tumors [61].

For patients with prostate cancer who experience disease relapse after local therapy or who have metastatic disease, androgen deprivation therapy (ADT) remains the cornerstone of systemic treatment. However, despite significant initial responses to ADT, nearly all metastatic patients eventually progress to incurable metastatic castration-resistant prostate cancer (mCRPC) [69]. Using digital PCR, Tjon-Kon-Fat and colleagues analyzed TEPs to identify cancer-derived transcripts in mCRPC patients before initiating therapy. The study included 24 patients receiving docetaxel-based chemotherapy as first-line treatment; 26 patients treated at any stage of castration-resistant disease with abiraterone, an irreversible inhibitor of the cytochrome P450 17α-hydroxy/17,20-lyase (CYP17) enzyme responsible for androgen synthesis; and 15 healthy controls. Transcripts of prostate cancer-associated biomarkers—kallikrein-related peptidase-2 and -3 (KLK2, KLK3), folate hydrolase 1 (FOLH1), and neuropeptide Y (NPY)—were detected in the platelets of cancer patients but not in those of healthy controls. In the docetaxel-treated cohort, no biomarkers correlated with PFS, and only detectable FOLH1 levels were linked to shorter OS ($p < 0.001$). However, in the abiraterone-treated cohort, FOLH1 was associated with a poor PSA response ($p < 0.05$), while patients with detectable levels of KLK3, KLK2, and FOLH1 exhibited significantly increased risks of death, with hazard ratios of 4.7 ($p < 0.01$; 95% CI, 1.6–13.6) for KLK3, 5.3 ($p < 0.001$; 95% CI, 1.9–14.5) for KLK2, and 3.0 ($p < 0.05$; 95% CI, 1.1–8.1) for FOLH1. Moreover, KLK3 (HR 2.6, $p < 0.01$; 95% CI, 1.1–6.2), NPY (HR 2.7, $p < 0.05$; 95% CI, 1.0–6.8), and FOLH1 (HR 3.4, $p < 0.05$; 95% CI, 1.2–9.8) were all associated with shorter PFS. Notably, a three-gene panel comprising KLK3, FOLH1, and NPY effectively distinguished responders from non-responders to abiraterone therapy, with patients positive for all three biomarkers showing a 4.2-fold increased risk of therapy failure ($p < 0.01$) and the panel achieving 87% sensitivity and 82% specificity for identifying patients with PFS longer than 6.5 months, offering a robust tool to identify resistance to treatment [59].

In summary, TEP analysis in liquid biopsies shows promise as a non-invasive tool for molecular profiling and treatment monitoring in prostate cancer. While early detection remains challenging, the ThromboSeq algorithm and biomarker panels highlight the potential of TEPs in identifying tumor characteristics, predicting therapeutic outcomes, and enhancing personalized treatment strategies (Table 1).

2.3. TEPs and Kidney Cancer

Kidney cancer ranks as the 14th most common cancer globally across both sexes. By 2045, the incidence of kidney cancer is projected to reach 616,030 cases with an estimated 256,427 deaths [65]. RCC accounts for approximately 85% of kidney tumors, with clear cell histology comprising around 70% of RCC cases. The diagnostic gold standard remains histologic evaluation via surgical resection or biopsy. While 70% of RCC cases are diagnosed at a localized stage, the remaining 30% present with metastatic disease. Despite curative-intent treatments such as surgery or focal therapies, one-third of localized tumors recur during follow-up, significantly increasing the risk of mortality, although survival rates have improved, with localized RCC achieving a 5-year survival rate of 90% [70,71].

Although early detection is critical to further improving survival outcomes in RCC, unlike prostate cancer, where the PSA test is part of a standard population screening program, there is currently no equivalent screening method for RCC. Early detection thus remains a significant challenge, largely due to the low clinical manifestation of symptoms in localized stages, resulting in most diagnoses being incidental findings during radiological tests conducted for unrelated reasons. In recent years, there has been growing interest in the use of blood biomarkers and liquid biopsy for the diagnosis and management of RCC [72,73]. While TEPs remain among the least explored biomarkers in this context, the established prognostic value of platelets—whose count is already integrated into risk stratification models for metastatic RCC treated with VEGF-targeted agents [74] and is correlated with aggressive tumor features in localized RCC [75]—underscores the ongoing need for more extensive research in this field.

In this evolving landscape, Kidney Injury Molecule-1 (KIM-1) has emerged as a particularly promising candidate. It is a transmembrane glycoprotein encoded by the *HAVCR1* gene, notably overexpressed in clear cell and papillary RCC. Its soluble ectodomain, detectable in both plasma and serum, enhances malignancy prediction when measured preoperatively and serves as an indicator of minimal residual disease, recurrence risk following nephrectomy, and potential benefit from adjuvant immunotherapy, as highlighted in recent studies [76–78]. In 2020, Sol and colleagues employed RNA panels derived from TEPs, identified through the digital SWARM algorithm, which leverages swarm intelligence, to detect and monitor glioblastoma. Their study assessed the accuracy of detection by comparing spliced RNA profiles from glioblastoma patients with those of individuals diagnosed with multiple sclerosis or brain metastases and healthy controls. Intriguingly, among the patients with brain metastases, seven were identified as having RCC. However, the study did not include a detailed subgroup analysis specific to the RCC cohort, leaving a gap in understanding the applicability of this approach to RCC-related metastases [55].

In a study more specifically focused on RCC, Xiao and colleagues investigated the use of TEPs as a minimally invasive biomarker for detecting RCC. Using RNA sequencing (RNA-seq), they analyzed platelet RNA profiles from 24 RCC patients and 25 controls (including healthy donors and individuals with benign renal tumors) and identified 203 differentially expressed genes, with 64 upregulated and 139 downregulated in RCC patients. With machine learning techniques, they developed a 68-gene TEP-based diagnostic model that showed high accuracy across datasets: 100% (AUC: 1.000) in the training set, 88.9% (AUC: 0.963) in the validation set, and 95.9% (AUC: 0.988) in the overall cohort. Despite certain limitations, including the small sample size and lack of external validation, these findings underscore the potential of TEPs as a robust and accurate blood-based biomarker platform, providing a non-invasive alternative for RCC detection and paving the way for further validation and clinical application [63].

In the study by In 't Veld and colleagues using ThromboSeq [61], the test achieved an AUC of 0.87 and a prediction accuracy of 66% in the sub-cohort of RCC patients. Furthermore, the test successfully identified the tumor origin in over 80% of cancer patients, including a category grouping urothelial, prostate, and renal tumors. Platelet RNA profiles were shown to be influenced by both primary tumors and metastases, particularly brain metastases. Nevertheless, although RCC was among the tumor types analyzed, detailed results specific to this subgroup were not reported [61].

In conclusion, TEP RNA analysis shows great potential as a non-invasive tool for RCC diagnosis, offering high accuracy for early detection and tumor characterization. However, its clinical use is limited by studies with small sample sizes, lack of validation, and incomplete mechanistic understanding (Table 1). Large-scale studies are needed to validate findings, optimize diagnostic models, and explore the role of TEPs in monitoring metastasis and treatment response, thereby paving the way for enhanced RCC management and broader applications.

2.4. TEPs and Bladder Cancer

Urinary bladder tumors are the 9th most common cancer overall and the 6th among men. Projections indicate a substantial rise in the incidence of bladder cancer over the next two decades, with cases potentially surpassing one million [65,79]. More than 90% of urothelial tumors originate in the urinary bladder, with urothelial carcinoma being the predominant histological subtype. It is essential to differentiate between non-muscle-invasive and muscle-invasive urothelial tumors, as their prognosis and management differ significantly. Approximately 75% of cases are non-muscle-invasive, which typically has a lower risk of recurrence and progression. As a result, aggressive therapeutic approaches are generally unnecessary for these patients. Instead, treatment focuses primarily on endoscopic interventions aimed at minimizing the risks of recurrence and progression [80,81].

Muscle-invasive bladder cancer (MIBC) has a global 5-year mortality rate of approximately 40–50% and poses a significant clinical challenge due to its aggressive biological behavior. Treatment decisions for MIBC are predominantly based on histopathological evaluation of biopsies from primary tumors and clinical staging, which together guide the therapeutic approach [82]. The standard treatment for patients with MIBC involves neoadjuvant cisplatin-based chemotherapy followed by radical cystectomy, which offers the best chance for a curative outcome. Recently, combining chemotherapy with immunotherapy has also demonstrated improved outcomes [83]. Nevertheless, despite this aggressive multimodal approach, metastatic recurrence remains frequent, and the 5-year survival rate is limited to 50–60% [84,85]. These suboptimal outcomes underscore the urgent need for reliable prognostic and predictive biomarkers to refine treatment strategies. Such biomarkers could help identify patients who may be able to avoid invasive procedures like cystectomy, thereby preserving their quality of life without compromising outcomes. Current research is actively evaluating novel blood-based biomarkers, such as CTCs, ctDNA, urine proteins, exosomes, and metabolites, which show potential for advancing personalized medicine in the management of MIBC [86,87].

The use of ctDNA as a predictive and prognostic biomarker represents one of the most advanced and clinically applicable approaches in this disease [88–91]. For example, the Phase 3 IMvigor010 trial evaluated the efficacy of one year of immune checkpoint inhibitor therapy with adjuvant atezolizumab, a PD-L1 inhibitor, following cystectomy in patients with high-risk MIBC. Although 809 patients were randomized, the study did not meet its primary endpoint of improving disease-free survival [92]. However, this trial also explored the role of ctDNA as a biomarker for predicting outcomes and guiding treatment decisions. Patients with detectable ctDNA had worse survival outcomes although they also derived benefit from atezolizumab, with significantly higher survival rates and evidence of ctDNA clearance or reduction [93,94]. Along the same lines, results from the KEYNOTE-361 study, presented at the 2024 ASCO Annual Meeting, showed that ctDNA levels act as a prognostic indicator for patients with advanced urothelial carcinoma receiving pembrolizumab, particularly in evaluating treatment response as compared to traditional chemotherapy [95]. These findings underscore the potential of ctDNA positivity and clearance as key indicators of clinical outcomes, supporting the adoption of a personalized approach for patient selection in postoperative cancer treatment.

Unlike ctDNA, the role of TEPs in bladder cancer has not been extensively investigated. However, Hinsenveld and colleagues designed a prospective study aimed at evaluating the potential of predicting a complete pathological response in patients with MIBC undergoing neoadjuvant chemotherapy. The ultimate goal is to identify patients who may safely avoid radical cystectomy. Their approach involves integrating clinical–radiological, histological, and molecular biomarker parameters derived from tissue, urine, and blood samples, including the exploration of TEPs as part of this multimodal assessment [64].

In the study by In 't Veld and colleagues, the pan-cancer ThromboSeq assay achieved an 89% accuracy rate and an AUC of 0.99 in patients with urothelial cancer and accurately detected the tumor origin in over 80% of cases within a mixed cohort [61].

In conclusion, while emerging biomarkers such as ctDNA have shown substantial promise in predicting prognosis and guiding treatment strategies in bladder cancer, the potential role of TEPs remains largely unexplored in this context, highlighting the need for more comprehensive studies to evaluate the usefulness of TEPs as predictive and prognostic biomarkers (Table 1).

3. Conclusions and Future Perspectives

TEPs are a promising source of tumor molecular data. They are easily accessible by liquid biopsy and offer unique advantages due to their abundance, ease of isolation, and dynamic RNA profiles that reflect tumor activity in real time. Current evidence underscores their potential as diagnostic, prognostic, and predictive biomarkers across a variety of solid tumors, including urological cancers. Unlike traditional tumor tissue biopsies, which are invasive and often fail to capture tumor heterogeneity, TEPs in liquid biopsies provide a minimally invasive alternative capable of delivering comprehensive molecular insights that can guide tailored treatment strategies to improve patient outcomes.

TEPs offer several distinct advantages over other blood-based biosources like ctDNA. While ctDNA can be fragmented and is often present at low concentrations, particularly in early-stage cancers or certain tumor types, TEPs are abundant and can be isolated using standardized techniques. Their ability to internalize RNA and other tumor-derived molecules provides a dynamic profile of the tumor molecular landscape. Moreover, they capture a broader range of signals, including RNA transcripts and splicing variants, which may not be detectable by ctDNA analysis. Importantly, TEPs reflect both primary and metastatic tumor activity, making them a versatile tool for cancer monitoring and guiding treatment decisions. Nonetheless, several technical challenges remain before TEP analysis can be widely adopted in clinical practice. Reproducibility is a major concern, and the lack of standardized protocols for platelet isolation, RNA extraction, and downstream bioinformatic analysis may lead to inconsistent results across studies. Although the team led by Thomas Wurdinger has proposed a robust and reproducible method for platelet isolation from whole blood for sequencing-based studies [42,96], its widespread adoption and validation across independent cohorts and clinical settings is still pending. Another key challenge lies in minimizing molecular noise, which may arise from platelet interactions with non-cancerous cells or systemic inflammatory processes. One strategy to address this issue involves parallel analysis of gene expression in tumor tissue to validate that the biomarkers identified in TEPs truly originate from the tumor and are not confounded by other physio pathological conditions. This comparative approach has been employed in several studies [53,60], reinforcing the tumor-specific nature of selected TEP-derived RNA signatures. Addressing these methodological limitations is critical for enhancing the reliability, sensitivity, and specificity of TEP-based assays, ultimately paving the way for their integration into clinical decision-making workflows.

In urological tumors, the role of TEPs is particularly compelling though underexplored. In prostate cancer, emerging evidence highlights the potential of TEP RNA profiles to detect early disease and predict therapeutic responses, although further validation in prospective, multicenter studies is required. In kidney cancer, preliminary studies have demonstrated high diagnostic accuracy, but the clinical relevance of these findings needs to be evaluated in larger cohorts. Notably, given the current challenges in early detection and the lack of effective screening strategies for RCC, TEP analysis could represent a valuable non-invasive diagnostic approach, potentially enabling earlier intervention and improved outcomes. In bladder cancer, the exploration of TEPs remains limited but with a potential application for monitoring treatment response and guiding bladder preservation strategies. Future research should focus on integrating TEP analysis into multimodal biomarker panels to enhance its predictive power, particularly in MIBC patients undergoing neoadjuvant chemotherapy. This approach could reduce treatment delays, prevent toxicities in non-responsive patients, and optimize the timing of surgery. It is also worth noting that testicular germ cell tumors and penile cancer have not been included in this review, as—likely due to their low incidence—no studies have yet investigated the potential prognostic or predictive role of TEPs in these malignancies. This represents a gap in current knowledge that deserves to be explored in future research.

Despite the promising potential of TEPs, several important questions remain unanswered. One key area is the longitudinal stability of TEP RNA signatures over the course of disease progression and treatment—understanding whether these profiles remain consistent or evolve in response to therapeutic pressure could enhance their value in monitoring. Additionally, it is still unclear to what extent TEPs are able to capture intratumoral heterogeneity or reflect emerging resistance mechanisms in real time. Further research is also needed to determine how TEP profiles vary among different subtypes and stages of urological cancers and whether they can be tailored to specific clinical scenarios. Finally, the cost effectiveness, scalability, and integration of TEP-based assays into existing diagnostic and therapeutic workflows remain open challenges that must be addressed before widespread clinical implementation can be achieved.

In conclusion, while significant progress has been made, the clinical integration of TEPs as blood-based biomarkers requires robust validation and technological refinement. Expanding research efforts in this field will be crucial to unlocking the full potential of TEPs and transforming their promise into tangible benefits for patients, particularly in the realm of urological tumors.

Author Contributions: M.F., acquisition of data, methodology, analysis and interpretation of data, visualization, writing—original draft, and writing—review and editing; S.C., acquisition of data, methodology, analysis and interpretation of data, and writing—original draft; M.F.-S., acquisition of data, methodology, analysis and interpretation of data, and writing—original draft; I.P.-C., writing—review and editing; A.B., writing—review and editing; A.F., writing—review and editing and supervision; V.R.d.P., conceptualization, acquisition of data, methodology, analysis and interpretation of data, writing—original draft, writing—review and editing, visualization, project administration, and supervision. All authors have read and agreed to the published version of the manuscript.

Funding: This research received no external funding.

Institutional Review Board Statement: Not applicable.

Informed Consent Statement: Not applicable.

Data Availability Statement: Not applicable.

Acknowledgments: The authors thank Renée Grupp for assistance in drafting the manuscript.

Conflicts of Interest: The authors declare no conflicts of interest.

Abbreviations

The following abbreviations are used in this manuscript:

ALK	Anaplastic Lymphoma Kinase
AUC	Area Under the Curve
bFGF	Basic Fibroblast Growth Factor
CD40L	CD40 Ligand
CTCs	Circulating Tumor Cells
ctDNA	Circulating Tumor DNA
EGF	Epidermal Growth Factor
EGFR	Epidermal Growth Factor Receptor
EGFRvIII	Epidermal Growth Factor Receptor Variant III
EML4	Echinoderm Microtubule Associated Protein-Like 4
EMT	Epithelial–Mesenchymal Transition

EVs	Extracellular Vesicles
EZH2	Enhancer of Zeste Homolog 2
GARP	TGFβ Docking Receptor Glycoprotein A Repetitions Predominant
G-CSF	Granulocyte Colony Stimulating Factor
GM-CSF	Granulocyte Macrophage-Colony Stimulating Factor
GPIIb/IIIa	Glycoprotein IIb/IIIa
IL-6	Interleukin-6
KRAS	Kirsten Rat Sarcoma Virus
MALAT1	Metastasis Associated Lung Adenocarcinoma Transcript 1
MDSCs	Myeloid-Derived Suppressor Cells
MHC	Major Histocompatibility Complex
MIBC	Muscle-Invasive Bladder Cancer
mCRPC	Metastatic Castration-Resistant Prostate Cancer
MMPs	Metalloproteinases
NK Cells	Natural Killer Cells
NSCLC	Non-Small Cell Lung Cancer
OS	Overall Survival
PCA3	Prostate Cancer Antigen 3
PDGF	Platelet-Derived Growth Factor
PD-L1	Programmed Death Ligand 1
PIK3CA	Phosphatidylinositol-4,5-Bisphosphate 3-Kinase, Catalytic Subunit Alpha
PFS	Progression-Free Survival
PSA	Prostate-Specific Antigen
PSMA	Prostate-Specific Membrane Antigen
RCC	Renal Cell Carcinoma
TAMs	Tumor-Associated Macrophages
TEPs	Tumor-Educated Platelets
TGF-β	Transforming Growth Factor-Beta
TIMP1	TIMP Metallopeptidase Inhibitor 1
TME	Tumor Microenvironment
Tregs	Regulatory T Cells
TSP-1	Thrombospondin-1
VEGF	Vascular Endothelial Growth Factor

References

1. Menter, D.G.; Tucker, S.C.; Kopetz, S.; Sood, A.K.; Crissman, J.D.; Honn, K.V. Platelets and cancer: A casual or causal relationship: Revisited. *Cancer Metastasis Rev.* **2014**, *33*, 231–269. [PubMed]
2. Jain, S.; Harris, J.; Ware, J. Platelets: Linking hemostasis and cancer. *Arter. Thromb. Vasc. Biol.* **2010**, *30*, 2362–2367.
3. Connolly, G.C.; Phipps, R.P.; Francis, C.W. Platelets and cancer-associated thrombosis. *Semin. Oncol.* **2014**, *41*, 302–310. [PubMed]
4. Li, S.; Lu, Z.; Wu, S.; Chu, T.; Li, B.; Qi, F.; Zhao, Y.; Nie, G. The dynamic role of platelets in cancer progression and their therapeutic implications. *Nat. Rev. Cancer* **2024**, *24*, 72–87.
5. Pinedo, H.M.; Verheul, H.M.; D'Amato, R.J.; Folkman, J. Involvement of platelets in tumour angiogenesis? *Lancet* **1998**, *352*, 1775–1777.
6. Baj-Krzyworzeka, M.; Majka, M.; Pratico, D.; Ratajczak, J.; Vilaire, G.; Kijowski, J.; Reca, R.; Janowska-Wieczorek, A.; Ratajczak, M.Z. Platelet-derived microparticles stimulate proliferation, survival, adhesion, and chemotaxis of hematopoietic cells. *Exp. Hematol.* **2002**, *30*, 450–459.
7. Labelle, M.; Begum, S.; Hynes, R.O. Direct signaling between platelets and cancer cells induces an epithelial-mesenchymal-like transition and promotes metastasis. *Cancer Cell* **2011**, *20*, 576–590.
8. Bothina, S.M.I.W.B.; Iman, M.; Maha, M.S.; Mamdouh, R.; Faiza, M.E. Platelet and neutrophil cross-talk-mediating cancer growth and metastasis in patients with urinary bladder cancer. *J. Arab. Soc. Med. Res.* **2012**, *7*, 1–9.
9. Zhou, L.; Zhang, Z.; Tian, Y.; Li, Z.; Liu, Z.; Zhu, S. The critical role of platelet in cancer progression and metastasis. *Eur. J. Med. Res.* **2023**, *28*, 385.

10. Nierodzik, M.L.; Karpatkin, S. Thrombin induces tumor growth, metastasis, and angiogenesis: Evidence for a thrombin-regulated dormant tumor phenotype. *Cancer Cell* **2006**, *10*, 355–362.
11. Hanahan, D.; Weinberg, R.A. Hallmarks of cancer: The next generation. *Cell* **2011**, *144*, 646–674.
12. Klement, G.L.; Yip, T.T.; Cassiola, F.; Kikuchi, L.; Cervi, D.; Podust, V.; Italiano, J.E.; Wheatley, E.; Abou-Slaybi, A.; Bender, E.; et al. Platelets actively sequester angiogenesis regulators. *Blood* **2009**, *113*, 2835–2842. [CrossRef] [PubMed]
13. Sabrkhany, S.; Griffioen, A.W.; Oude Egbrink, M.G. The role of blood platelets in tumor angiogenesis. *Biochim. Biophys. Acta* **2011**, *1815*, 189–196.
14. Wojtukiewicz, M.Z.; Sierko, E.; Hempel, D.; Tucker, S.C.; Honn, K.V. Platelets and cancer angiogenesis nexus. *Cancer Metastasis Rev.* **2017**, *36*, 249–262. [PubMed]
15. O'Byrne, K.J.; Dobbs, N. Vascular endothelial growth factor platelet counts, and prognosis in renal cancer. *Lancet* **1999**, *353*, 1494–1495.
16. Cumpanas, A.A.; Cimpean, A.M.; Ferician, O.; Ceausu, R.A.; Sarb, S.; Barbos, V.; Dema, A.; Raica, M. The Involvement of PDGF-B/PDGFRbeta Axis in the Resistance to Antiangiogenic and Antivascular Therapy in Renal Cancer. *Anticancer Res.* **2016**, *36*, 2291–2295. [PubMed]
17. Guillaume, Z.; Auvray, M.; Vano, Y.; Oudard, S.; Helley, D.; Mauge, L. Renal Carcinoma and Angiogenesis: Therapeutic Target and Biomarkers of Response in Current Therapies. *Cancers* **2022**, *14*, 6167. [CrossRef]
18. Joyce, J.A.; Pollard, J.W. Microenvironmental regulation of metastasis. *Nat. Rev. Cancer* **2009**, *9*, 239–252.
19. Fu, L.Q.; Du, W.L.; Cai, M.H.; Yao, J.Y.; Zhao, Y.Y.; Mou, X.Z. The roles of tumor-associated macrophages in tumor angiogenesis and metastasis. *Cell Immunol.* **2020**, *353*, 104119.
20. Li, Y.; Wang, H.; Zhao, Z.; Yang, Y.; Meng, Z.; Qin, L. Effects of the interactions between platelets with other cells in tumor growth and progression. *Front. Immunol.* **2023**, *14*, 1165989.
21. Tuerhong, N.; Yang, Y.; Wang, C.; Huang, P.; Li, Q. Interactions between platelets and the cancer immune microenvironment. *Crit. Rev. Oncol. Hematol.* **2024**, *199*, 104380. [PubMed]
22. Okazaki, M.; Yamaguchi, T.; Tajima, H.; Fushida, S.; Ohta, T. Platelet adherence to cancer cells promotes escape from innate immune surveillance in cancer metastasis. *Int. J. Oncol.* **2020**, *57*, 980–988. [PubMed]
23. Placke, T.; Kopp, H.G.; Salih, H.R. The wolf in sheep's clothing: Platelet-derived "pseudo self" impairs cancer cell "missing self" recognition by NK cells. *Oncoimmunology* **2012**, *1*, 557–559. [CrossRef] [PubMed]
24. Wu, X.; Li, T.; Jiang, R.; Yang, X.; Guo, H.; Yang, R. Targeting MHC-I molecules for cancer: Function, mechanism, and therapeutic prospects. *Mol. Cancer* **2023**, *22*, 194.
25. Zaslavsky, A.B.; Adams, M.P.; Cao, X.; Maj, T.; Choi, J.E.; Stangl-Kremser, J.; Patel, S.; Putelo, A.; Lee, S.K.; Nallandhighal, S.; et al. Platelet PD-L1 suppresses anti-cancer immune cell activity in PD-L1 negative tumors. *Sci. Rep.* **2020**, *10*, 19296.
26. Lutz, M.S.; Klimovich, B.; Maurer, S.; Heitmann, J.S.; Marklin, M.; Zekri, L.; Jung, G.; Salih, H.R.; Hinterleitner, C. Platelets subvert antitumor efficacy of T cell-recruiting bispecific antibodies. *J. Immunother. Cancer* **2022**, *10*, e003655. [CrossRef]
27. Kopp, H.G.; Placke, T.; Salih, H.R. Platelet-derived transforming growth factor-beta down-regulates NKG2D thereby inhibiting natural killer cell antitumor reactivity. *Cancer Res.* **2009**, *69*, 7775–7783. [CrossRef]
28. Tran, D.Q.; Andersson, J.; Wang, R.; Ramsey, H.; Unutmaz, D.; Shevach, E.M. GARP (LRRC32) is essential for the surface expression of latent TGF-beta on platelets and activated FOXP3+ regulatory T cells. *Proc. Natl. Acad. Sci. USA* **2009**, *106*, 13445–13450. [CrossRef]
29. Rachidi, S.; Metelli, A.; Riesenberg, B.; Wu, B.X.; Nelson, M.H.; Wallace, C.; Paulos, C.M.; Rubinstein, M.P.; Garrett-Mayer, E.; Hennig, M.; et al. Platelets subvert T cell immunity against cancer via GARP-TGFbeta axis. *Sci. Immunol.* **2017**, *2*, eaai7911. [CrossRef]
30. Zimmer, N.; Krebs, F.K.; Zimmer, S.; Mitzel-Rink, H.; Kumm, E.J.; Jurk, K.; Grabbe, S.; Loquai, C.; Tuettenberg, A. Platelet-Derived GARP Induces Peripheral Regulatory T Cells-Potential Impact on T Cell Suppression in Patients with Melanoma-Associated Thrombocytosis. *Cancers* **2020**, *12*, 3653. [CrossRef]
31. Servais, L.; Wera, O.; Epoh, J.D.; Delierneux, C.; Bouznad, N.; Rahmouni, S.; Mazzucchelli, G.; Baiwir, D.; Delvenne, P.; Lancellotti, P.; et al. Platelets contribute to the initiation of colitis-associated cancer by promoting immunosuppression. *J. Thromb. Haemost.* **2018**, *16*, 762–777. [CrossRef] [PubMed]
32. Hatogai, K.; Sweis, R.F. The Tumor Microenvironment of Bladder Cancer. *Adv. Exp. Med. Biol.* **2020**, *1296*, 275–290. [PubMed]
33. Welch, D.R.; Hurst, D.R. Defining the Hallmarks of Metastasis. *Cancer Res.* **2019**, *79*, 3011–3027. [CrossRef] [PubMed]
34. Schlesinger, M. Role of platelets and platelet receptors in cancer metastasis. *J. Hematol. Oncol.* **2018**, *11*, 125. [CrossRef]
35. Liao, K.; Zhang, X.; Liu, J.; Teng, F.; He, Y.; Cheng, J.; Yang, Q.; Zhang, W.; Xie, Y.; Guo, D.; et al. The role of platelets in the regulation of tumor growth and metastasis: The mechanisms and targeted therapy. *MedComm* **2023**, *4*, e350. [CrossRef]

36. Ward, Y.; Lake, R.; Faraji, F.; Sperger, J.; Martin, P.; Gilliard, C.; Ku, K.P.; Rodems, T.; Niles, D.; Tillman, H.; et al. Platelets Promote Metastasis via Binding Tumor CD97 Leading to Bidirectional Signaling that Coordinates Transendothelial Migration. *Cell Rep.* **2018**, *23*, 808–822. [CrossRef]
37. Mezei, T.; Bode, I.; Tenke, P.; Josa, V.; Merkel, K.; Szilasi, Z.; Tordai, A.; Mathe, D.; Baranyai, Z. The Correlation Between Platelet Count and Survival in Prostate Cancer. *Res. Rep. Urol.* **2022**, *14*, 193–202. [CrossRef]
38. Garofano, K.; Rashid, K.; Smith, M.; Brantner, C.; Suwunnakorn, S.; Diemert, D.; Gordon, O.; Horvath, A.; Khan, S.; Popratiloff, A.; et al. Prostate cancer cell-platelet bidirectional signaling promotes calcium mobilization, invasion and apoptotic resistance via distinct receptor-ligand pairs. *Sci. Rep.* **2023**, *13*, 2864. [CrossRef]
39. Kessenbrock, K.; Plaks, V.; Werb, Z. Matrix metalloproteinases: Regulators of the tumor microenvironment. *Cell* **2010**, *141*, 52–67. [CrossRef]
40. Lee, T.; Esemuede, N.; Sumpio, B.E.; Gahtan, V. Thrombospondin-1 induces matrix metalloproteinase-2 activation in vascular smooth muscle cells. *J. Vasc. Surg.* **2003**, *38*, 147–154. [CrossRef]
41. McLaren, K.M. Immunohistochemical localisation of thrombospondin in human megakaryocytes and platelets. *J. Clin. Pathol.* **1983**, *36*, 197–199. [PubMed]
42. Best, M.G.; Sol, N.; Kooi, I.; Tannous, J.; Westerman, B.A.; Rustenburg, F.; Schellen, P.; Verschueren, H.; Post, E.; Koster, J.; et al. RNA-Seq of Tumor-Educated Platelets Enables Blood-Based Pan-Cancer, Multiclass, and Molecular Pathway Cancer Diagnostics. *Cancer Cell* **2015**, *28*, 666–676. [PubMed]
43. Najafi, S.; Asemani, Y.; Majidpoor, J.; Mahmoudi, R.; Aghaei-Zarch, S.M.; Mortezaee, K. Tumor-educated platelets. *Clin. Chim. Acta* **2024**, *552*, 117690. [PubMed]
44. Roweth, H.G.; Battinelli, E.M. Lessons to learn from tumor-educated platelets. *Blood* **2021**, *137*, 3174–3180.
45. D'Ambrosi, S.; Nilsson, R.J.; Wurdinger, T. Platelets and tumor-associated RNA transfer. *Blood* **2021**, *137*, 3181–3191.
46. Varkey, J.; Nicolaides, T. Tumor-Educated Platelets: A Review of Current and Potential Applications in Solid Tumors. *Cureus* **2021**, *13*, e19189.
47. Wang, Y.; Zhang, H.; Li, H.; Xiong, J.; Wang, J.; Huang, Y. Application of tumor-educated platelets as new fluid biopsy markers in various tumors. *Clin. Transl. Oncol.* **2023**, *25*, 114–125.
48. Wurdinger, T.; In 't Veld, S.; Best, M.G. Platelet RNA as Pan-Tumor Biomarker for Cancer Detection. *Cancer Res.* **2020**, *80*, 1371–1373.
49. Joosse, S.A.; Pantel, K. Tumor-Educated Platelets as Liquid Biopsy in Cancer Patients. *Cancer Cell* **2015**, *28*, 552–554.
50. Connal, S.; Cameron, J.M.; Sala, A.; Brennan, P.M.; Palmer, D.S.; Palmer, J.D.; Perlow, H.; Baker, M.J. Liquid biopsies: The future of cancer early detection. *J. Transl. Med.* **2023**, *21*, 118.
51. In 't Veld, S.; Wurdinger, T. Tumor-educated platelets. *Blood* **2019**, *133*, 2359–2364. [PubMed]
52. Calverley, D.C.; Phang, T.L.; Choudhury, Q.G.; Gao, B.; Oton, A.B.; Weyant, M.J.; Geraci, M.W. Significant downregulation of platelet gene expression in metastatic lung cancer. *Clin. Transl. Sci.* **2010**, *3*, 227–232. [PubMed]
53. Nilsson, R.J.A.; Balaj, L.; Hulleman, E.; van Rijn, S.; Pegtel, D.M.; Walraven, M.; Widmark, A.; Gerritsen, W.R.; Verheul, H.M.; Vandertop, W.P.; et al. Blood platelets contain tumor-derived RNA biomarkers. *Blood* **2011**, *118*, 3680–3683. [PubMed]
54. Nilsson, R.J.A.; Karachaliou, N.; Berenguer, J.; Gimenez-Capitan, A.; Schellen, P.; Teixido, C.; Tannous, J.; Kuiper, J.L.; Drees, E.; Grabowska, M.; et al. Rearranged EML4-ALK fusion transcripts sequester in circulating blood platelets and enable blood-based crizotinib response monitoring in non-small-cell lung cancer. *Oncotarget* **2016**, *7*, 1066–1075.
55. Sol, N.; In 't Veld, S.; Vancura, A.; Tjerkstra, M.; Leurs, C.; Rustenburg, F.; Schellen, P.; Verschueren, H.; Post, E.; Zwaan, K.; et al. Tumor-Educated Platelet RNA for the Detection and (Pseudo) progression Monitoring of Glioblastoma. *Cell Rep. Med.* **2020**, *1*, 100101.
56. Heinhuis, K.M.; In 't Veld, S.; Dwarshuis, G.; van den Broek, D.; Sol, N.; Best, M.G.; Coevorden, F.V.; Haas, R.L.; Beijnen, J.H.; van Houdt, W.J.; et al. RNA-Sequencing of Tumor-Educated Platelets, a Novel Biomarker for Blood-Based Sarcoma Diagnostics. *Cancers* **2020**, *12*, 1372. [CrossRef]
57. Wang, H.; Wei, X.; Wu, B.; Su, J.; Tan, W.; Yang, K. Tumor-educated platelet miR-34c-3p and miR-18a-5p as potential liquid biopsy biomarkers for nasopharyngeal carcinoma diagnosis. *Cancer Manag. Res.* **2019**, *11*, 3351–3360.
58. Yang, L.; Jiang, Q.; Li, D.Z.; Zhou, X.; Yu, D.S.; Zhong, J. TIMP1 mRNA in tumor-educated platelets is diagnostic biomarker for colorectal cancer. *Aging* **2019**, *11*, 8998–9012.
59. Hu, Z.; Wang, N.; Zhang, Y.; Zhang, D.; Sun, S.; Yu, H.; Lin, Y.; Zhao, X.; Wang, H.; Wu, X.; et al. PD-L1 mRNA derived from tumor-educated platelets as a potential immunotherapy biomarker in non-small cell lung cancer. *Transl. Lung Cancer Res.* **2024**, *13*, 345–354.
60. Hanze, J.; Jakubowski, P.; Heers, H.; Hegele, A.; Timmesfeld, N.; Hofmann, R.; Olbert, P.J. Assessing blood platelets as RNA biomarker source for prostate cancer. *Biomarkers* **2016**, *21*, 653–659.

61. GJG, S.; Arkani, M.; Post, E.; Antunes-Ferreira, M.; D'Ambrosi, S.; Vessies, D.C.; Vermunt, L.; Vancura, A.; Muller, M.; Niemeijer, A.L.N.; et al. Detection and localization of early- and late-stage cancers using platelet RNA. *Cancer Cell* **2022**, *40*, 999–1009.e6.
62. Tjon-Kon-Fat, L.A.; Lundholm, M.; Schroder, M.; Wurdinger, T.; Thellenberg-Karlsson, C.; Widmark, A.; Wikstrom, P.; Nilsson, R.J.A. Platelets harbor prostate cancer biomarkers and the ability to predict therapeutic response to abiraterone in castration resistant patients. *Prostate* **2018**, *78*, 48–53. [PubMed]
63. Xiao, R.; Liu, C.; Zhang, B.; Ma, L. Tumor-Educated Platelets as a Promising Biomarker for Blood-Based Detection of Renal Cell Carcinoma. *Front. Oncol.* **2022**, *12*, 844520.
64. Hinsenveld, F.J.; Noordman, B.J.; Boormans, J.L.; Voortman, J.; van Leenders, G.J.L.H.; van der Pas, S.L.; van Beek, S.C.; Oprea-Lager, D.E.; Vis, A.N. Prediction of pathological response following neoadjuvant chemotherapy in patients with muscle-invasive bladder cancer: The PRE-PREVENCYS trial. *BMC Cancer* **2021**, *21*, 1161.
65. Bray, F.; Laversanne, M.; Sung, H.; Ferlay, J.; Siegel, R.L.; Soerjomataram, I.; Jemal, A. Global cancer statistics 2022: GLOBOCAN estimates of incidence and mortality worldwide for 36 cancers in 185 countries. *CA A Cancer J. Clin.* **2024**, *74*, 229–263.
66. Sartor, O.; de Bono, J.S. Metastatic Prostate Cancer. *N. Engl. J. Med.* **2018**, *378*, 645–657.
67. Miquelestorena-Standley, E.; Jourdan, M.L.; Collin, C.; Bouvier, C.; Larousserie, F.; Aubert, S.; Gomez-Brouchet, A.; Guinebretiere, J.M.; Tallegas, M.; Brulin, B.; et al. Effect of decalcification protocols on immunohistochemistry and molecular analyses of bone samples. *Mod. Pathol.* **2020**, *33*, 1505–1517.
68. Boerrigter, E.; Groen, L.N.; Van Erp, N.P.; Verhaegh, G.W.; Schalken, J.A. Clinical utility of emerging biomarkers in prostate cancer liquid biopsies. *Expert. Rev. Mol. Diagn.* **2020**, *20*, 219–230.
69. Davies, A.; Conteduca, V.; Zoubeidi, A.; Beltran, H. Biological Evolution of Castration-resistant Prostate Cancer. *Eur. Urol. Focus* **2019**, *5*, 147–154.
70. Rose, T.L.; Kim, W.Y. Renal Cell Carcinoma: A Review. *JAMA* **2024**, *332*, 1001–1010.
71. Young, M.; Jackson-Spence, F.; Beltran, L.; Day, E.; Suarez, C.; Bex, A.; Powles, T.; Szabados, B. Renal cell carcinoma. *Lancet* **2024**, *404*, 476–491. [PubMed]
72. Zieren, R.C.; Zondervan, P.J.; Pienta, K.J.; Bex, A.; de Reijke, T.M.; Bins, A.D. Diagnostic liquid biopsy biomarkers in renal cell cancer. *Nat. Rev. Urol.* **2024**, *21*, 133–157. [PubMed]
73. Machaalani, M.; Eid, M.; Semaan, K.; El Hajj Chehade, R.; Nawfal, R.; Baca, S.C.; Choueiri, T.K. Liquid biopsy in renal cell carcinoma. *Oncologist* **2024**, *29*, 821–823. [PubMed]
74. Heng, D.Y.; Xie, W.; Regan, M.M.; Warren, M.A.; Golshayan, A.R.; Sahi, C.; Eigl, B.J.; Ruether, J.D.; Cheng, T.; North, S.; et al. Prognostic factors for overall survival in patients with metastatic renal cell carcinoma treated with vascular endothelial growth factor-targeted agents: Results from a large, multicenter study. *J. Clin. Oncol.* **2009**, *27*, 5794–5799. [CrossRef]
75. Sole, K. Thrombocytosis predicts mortality in patients with renal cell carcinoma. *Nat. Clin. Pract. Urol.* **2006**, *3*, 240. [CrossRef]
76. Xu, W.; Gaborieau, V.; Niman, S.M.; Mukeria, A.; Liu, X.; Maremanda, K.P.; Takakura, A.; Zaridze, D.; Freedman, M.L.; Xie, W.; et al. Plasma Kidney Injury Molecule-1 for Preoperative Prediction of Renal Cell Carcinoma Versus Benign Renal Masses, and Association With Clinical Outcomes. *J. Clin. Oncol.* **2024**, *42*, 2691–2701. [CrossRef]
77. Xu, W.; Puligandla, M.; Halbert, B.; Haas, N.B.; Flaherty, K.T.; Uzzo, R.G.; Dutcher, J.P.; DiPaola, R.S.; Sabbisetti, V.; Bhatt, R.S. Plasma KIM-1 Is Associated with Recurrence Risk after Nephrectomy for Localized Renal Cell Carcinoma: A Trial of the ECOG-ACRIN Research Group (E2805). *Clin. Cancer Res.* **2021**, *27*, 3397–3403. [CrossRef]
78. Albiges, L.; Bex, A.; Suárez, C.; Uzzo, R.; Tang, X.; Assaf, Z.J.; Dubey, S.; Goluboff, E.; Carter, C.; Banchereau, R.; et al. Circulating kidney injury molecule-1 (KIM-1) biomarker analysis in IMmotion010: A randomized phase 3 study of adjuvant (adj) atezolizumab (atezo) vs placebo (pbo) in patients (pts) with renal cell carcinoma (RCC) at increased risk of recurrence after resection. *J. Clin. Oncol.* **2024**, *42*, 4506.
79. Jubber, I.; Ong, S.; Bukavina, L.; Black, P.C.; Comperat, E.; Kamat, A.M.; Kiemeney, L.; Lawrentschuk, N.; Lerner, S.P.; Meeks, J.J.; et al. Epidemiology of Bladder Cancer in 2023: A Systematic Review of Risk Factors. *Eur. Urol.* **2023**, *84*, 176–190.
80. Babjuk, M.; Burger, M.; Compérat, E.M.; Gontero, P.; Mostafid, A.H.; Palou, J.; van Rhijn, B.W.; Rouprêt, M.; Shariat, S.F.; Sylvester, R.; et al. European Association of Urology Guidelines on Non-muscle-invasive Bladder Cancer (Ta, T1, and Carcinoma in Situ). *Eur. Urol.* **2022**, *81*, 75–94.
81. Kamat, A.M.; Hahn, N.M.; Efstathiou, J.A.; Lerner, S.P.; Malmstrom, P.U.; Choi, W.; Guo, C.C.; Lotan, Y.; Kassouf, W. Bladder cancer. *Lancet* **2016**, *388*, 2796–2810. [CrossRef] [PubMed]
82. de Porras, V.R.; Pardo, J.C.; Etxaniz, O.; Font, A. Neoadjuvant therapy for muscle-invasive bladder cancer: Current clinical scenario, future perspectives, and unsolved questions. *Crit. Rev. Oncol. Hematol.* **2022**, *178*, 103795. [CrossRef] [PubMed]

83. Powles, T.; Catto, J.W.; Galsky, M.D.; Al-Ahmadie, H.; Meeks, J.J.; Nishiyama, H.; Vu, T.Q.; Antonuzzo, L.; Wiechno, P.; Atduev, V.; et al. Perioperative Durvalumab with Neoadjuvant Chemotherapy in Operable Bladder Cancer. *N. Engl. J. Med.* **2024**, *391*, 1773–1786. [CrossRef] [PubMed]
84. Grossman, H.B.; Natale, R.B.; Tangen, C.M.; Speights, V.O.; Vogelzang, N.J.; Trump, D.L.; deVere White, R.W.; Sarosdy, M.F.; Wood, D.P., Jr.; Raghavan, D.; et al. Neoadjuvant chemotherapy plus cystectomy compared with cystectomy alone for locally advanced bladder cancer. *N. Engl. J. Med.* **2003**, *349*, 859–866. [CrossRef]
85. Witjes, J.A.; Bruins, H.M.; Carrión, A.; Cathomas, R.; Compérat, E.; Efstathiou, J.A.; Fietkau, R.; Gakis, G.; Lorch, A.; Martini, A.; et al. European Association of Urology Guidelines on Muscle-invasive and Metastatic Bladder Cancer: Summary of the 2023 Guidelines. *Eur. Urol.* **2024**, *85*, 17–31. [CrossRef]
86. Crocetto, F.; Barone, B.; Ferro, M.; Busetto, G.M.; La Civita, E.; Buonerba, C.; Di Lorenzo, G.; Terracciano, D.; Schalken, J.A. Liquid biopsy in bladder cancer: State of the art and future perspectives. *Crit. Rev. Oncol. Hematol.* **2022**, *170*, 103577. [CrossRef]
87. Li, S.; Xin, K.; Pan, S.; Wang, Y.; Zheng, J.; Li, Z.; Liu, X.; Liu, B.; Xu, Z.; Chen, X. Blood-based liquid biopsy: Insights into early detection, prediction, and treatment monitoring of bladder cancer. *Cell. Mol. Biol. Lett.* **2023**, *28*, 28.
88. Crupi, E.; de Padua, T.C.; Marandino, L.; Raggi, D.; Dyrskjot, L.; Spiess, P.E.; Sonpavde, G.P.; Kamat, A.M.; Necchi, A. Circulating tumor DNA as a Predictive and Prognostic Biomarker in the Perioperative Treatment of Muscle-invasive Bladder Cancer: A Systematic Review. *Eur. Urol. Oncol.* **2024**, *7*, 44–52. [CrossRef]
89. Gao, X.; Qi, W.; Li, J.; Xia, Y.; Ding, P.; Guo, D.; Shi, B.; Jiang, X. Prognostic and predictive role of circulating tumor DNA detection in patients with muscle invasive bladder cancer: A systematic review and meta-analysis. *Cancer Cell Int.* **2025**, *25*, 75. [CrossRef]
90. Sakatani, T.; Sumiyoshi, T.; Kita, Y.; Takada, H.; Nakamura, K.; Hamada, A.; Murakami, K.; Sano, T.; Goto, T.; Sawada, A.; et al. Clinical Utility of Serial Circulating Tumor DNA Analysis as a Minimally Invasive Biomarker in Advanced Urothelial Cancer. *JCO Precis. Oncol.* **2025**, *9*, e2400472. [CrossRef]
91. Bellmunt, J.; Russell, B.M.; Szabados, B.; Valderrama, B.P.; Nadal, R. Current and Future Role of Circulating DNA in the Diagnosis and Management of Urothelial Carcinoma. *Am. Soc. Clin. Oncol. Educ. Book.* **2025**, *45*, e471912. [CrossRef] [PubMed]
92. Bellmunt, J.; Hussain, M.; Gschwend, J.E.; Albers, P.; Oudard, S.; Castellano, D.; Daneshmand, S.; Nishiyama, H.; Majchrowicz, M.; Degaonkar, V.; et al. Adjuvant atezolizumab versus observation in muscle-invasive urothelial carcinoma (IMvigor010): A multicentre, open-label, randomised, phase 3 trial. *Lancet Oncol.* **2021**, *22*, 525–537. [CrossRef] [PubMed]
93. Powles, T.; Assaf, Z.J.; Davarpanah, N.; Banchereau, R.; Szabados, B.E.; Yuen, K.C.; Grivas, P.; Hussain, M.; Oudard, S.; Gschwend, J.E.; et al. ctDNA guiding adjuvant immunotherapy in urothelial carcinoma. *Nature* **2021**, *595*, 432–437. [CrossRef] [PubMed]
94. Powles, T.; Assaf, Z.J.; Degaonkar, V.; Grivas, P.; Hussain, M.; Oudard, S.; Gschwend, J.E.; Albers, P.; Castellano, D.; Nishiyama, H.; et al. Updated Overall Survival by Circulating Tumor DNA Status from the Phase 3 IMvigor010 Trial: Adjuvant Atezolizumab Versus Observation in Muscle-invasive Urothelial Carcinoma. *Eur. Urol.* **2024**, *85*, 114–122. [CrossRef]
95. Powles, T.; Chang, W.Y.-H.; Yamamoto, Y.; Muñoz-Langa, J.; Reyes, F.; Peer, A.; Yu, E.Y.; Cohen, G.; Lorch, A.; Bavle, A.; et al. Quantitative circulating tumor DNA (ctDNA) assessment in patients (pts) with advanced urothelial carcinoma (UC) treated with pembrolizumab (pembro) or platinum-based chemotherapy (chemo) from the phase 3 KEYNOTE-361 trial. *J. Clin. Oncol.* **2024**, *42* (Suppl. 16), 4518. [CrossRef]
96. Best, M.G.; Sol, N.; In 't Veld, S.; Vancura, A.; Muller, M.; Niemeijer, A.N.; Fejes, A.V.; Tjon Kon Fat, L.A.; Huis In 't Veld, A.E.; Leurs, C.; et al. Swarm Intelligence-Enhanced Detection of Non-Small-Cell Lung Cancer Using Tumor-Educated Platelets. *Cancer Cell* **2017**, *32*, 238–252 e9. [CrossRef]

International Journal of
Molecular Sciences

MDPI

Article

GABBR2 as a Downstream Effector of the Androgen Receptor Induces Cisplatin Resistance in Bladder Cancer

Mohammad Amin Elahi Najafi [1,2,†], Masato Yasui [1,2,†], Yuki Teramoto [1,2], Tomoyuki Tatenuma [1,2], Guiyang Jiang [1,2] and Hiroshi Miyamoto [1,2,3,*]

[1] Department of Pathology and Laboratory Medicine, University of Rochester Medical Center, Rochester, NY 14642, USA; mohammadamin_elahinajafi@urmc.rochester.edu (M.A.E.N.); yabomabo@gmail.com (M.Y.); tera1980@kuhp.kyoto-u.ac.jp (Y.T.); tatenuma@yokohama-cu.ac.jp (T.T.); jianggy@hotmail.com (G.J.)

[2] James P. Wilmot Cancer Institute, University of Rochester Medical Center, Rochester, NY 14642, USA

[3] Department of Urology, University of Rochester Medical Center, Rochester, NY 14642, USA

* Correspondence: hiroshi_miyamoto@urmc.rochester.edu

† These authors contributed equally to this work.

Abstract: The precise molecular mechanisms responsible for resistance to cisplatin-based chemotherapy in patients with bladder cancer remain elusive, while we have indicated that androgen receptor (AR) activity in urothelial cancer is associated with its sensitivity. Our DNA microarray analysis in control vs. AR-knockdown bladder cancer sublines suggested that the expression of a GABA B receptor GABBR2 and AR was correlated. The present study aimed to determine the functional role of GABBR2 in modulating cisplatin sensitivity in bladder cancer. AR knockdown and dihydrotestosterone treatment considerably reduced and induced, respectively, GABBR2 expression, and the effect of dihydrotestosterone was at least partially restored by an antiandrogen hydroxyflutamide. A chromatin immunoprecipitation assay further revealed the binding of AR to the promoter region of *GABBR2* in bladder cancer cells. Meanwhile, *GABBR2* expression was significantly elevated in a cisplatin-resistant bladder cancer subline, compared with control cells. In AR-positive bladder cancer cells, knockdown of GABBR2 or treatment with a selective GABA B receptor antagonist, CGP46381, considerably enhanced the cytotoxic activity of cisplatin. However, no additional effect of CGP46381 on cisplatin-induced growth suppression was seen in GABBR2-knockdown cells. Moreover, in the absence of cisplatin, CGP46381 treatment and GABBR2 knockdown showed no significant changes in cell proliferation or migration. These findings suggest that GABBR2 represents a key downstream effector of AR signaling in inducing resistance to cisplatin treatment. Accordingly, inhibition of GABBR2 has the potential of being a means of chemosensitization, especially in patients with AR/GABBR2-positive bladder cancer.

Keywords: androgen receptor; chemoresistance; cisplatin; GABBR2; urothelial cancer

1. Introduction

Urothelial carcinoma in the urinary bladder has represented a commonly diagnosed malignancy, especially in men, and the global number of bladder cancer-related deaths has indeed increased from 165,100 in 2012 [1] to 212,536 in 2020 [2]. Particularly, muscle-invasive bladder cancer is often associated with metastatic disease and poor patient outcome (e.g., the overall 5-year survival rate of 8.3% [3]). Moreover, urothelial carcinoma also arises in the upper urinary tract but is much more often (e.g., 60% [4]) invasive at the initial diagnosis.

Despite the availability of novel therapeutic options, such as immunotherapy with checkpoint inhibitors, cisplatin (CDDP)-based chemotherapeutic regimens remain the first-line systemic treatment for locally advanced or metastatic urothelial carcinoma [5–8]. In addition, a meta-analysis involving 13,391 patients showed a significant overall survival benefit of neoadjuvant chemotherapy with CDDP-containing regimens prior to radical cystectomy [9]. Nonetheless, the response rate to CDDP therapy is not necessarily high (e.g., up to 60% [10–12]) in patients with bladder cancer. Accordingly, the development of not only novel anti-cancer agents but also strategies for chemosensitization of existing drugs should provide considerable improvement in oncologic outcomes of urothelial cancer.

The molecular mechanisms underlying CDDP resistance are still far from being fully understood [13–18]. Meanwhile, androgen-mediated androgen receptor (AR) signaling has been implicated in the promotion of urothelial tumorigenesis, which may be a reason for a substantially higher risk of bladder cancer development in men than in women, as well as urothelial tumor progression [19]. More fascinatingly, AR activity has also been associated with sensitivity to conventional nonsurgical treatments for bladder cancer [20], including CDDP-based chemotherapy [21–23]. Specifically, AR activation in bladder cancer cells has been shown to result in the induction of CDDP resistance, and AR knockdown or antiandrogen treatment thus enhances its sensitivity. Importantly, it remains to be further elucidated how the AR pathway modulates chemosensitivity in urothelial cancer. The present study aimed to explore whether gamma-aminobutyric acid (GABA) B receptor 2 (GABBR2), which belongs to the G protein-coupled receptor superfamily, functions as a downstream effector of AR and induces resistance to CDDP therapy in bladder cancer.

2. Results

2.1. AR Activity and GABBR2 Expression

We had recently employed DNA microarray analysis in control AR-positive bladder cancer UMUC3 cells versus a subline of UMUC3 stably expressing AR-shRNA [24]. We then investigated whether the expression of 25 candidate genes was correlated with that of AR. We eventually selected one of the candidates, GABBR2, and confirmed that its protein expression was indeed downregulated in AR-knockdown cells, compared with control cells (Figure 1A). A quantitative PCR in UMUC3 cells further showed that androgen (i.e., dihydrotestosterone (DHT)) treatment induced *GABBR2* expression, which was at least partially blocked by antiandrogen (i.e., hydroxyflutamide (HF)) treatment (Figure 1B). Thus, GABBR2 expression was correlated with the expression and activity of AR in bladder cancer cells. Meanwhile, the level of *GABBR2* expression in a CDDP-resistant subline was significantly higher than that in control cells (Figure 1C).

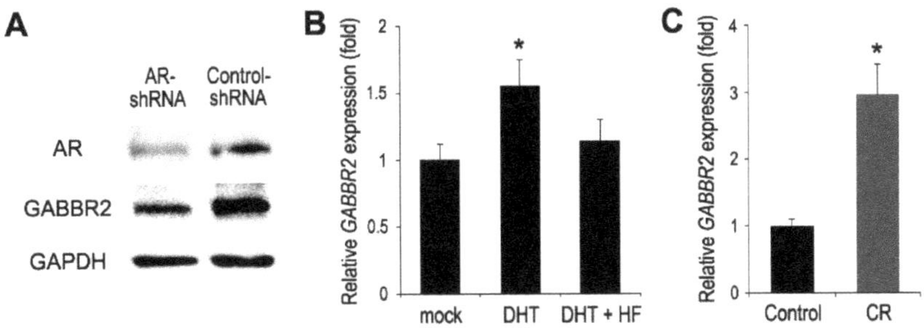

Figure 1. Associations between AR signaling and GABBR2 expression in bladder cancer cells. (**A**) Western blotting of AR and GABBR2 in UMUC3-AR-shRNA vs. UMUC3-control-shRNA. Real-time PCR of *GABBR2* in UMUC3 cultured for 24 h with ethanol (mock), 10 nM DHT, and/or 5 μM HF (**B**) or control UMUC3 vs. CDDP-resistant (CR) UMUC3 (**C**). The expression of *GABBR2* normalized to that of *GAPDH* and representing the mean (+SD) of at least three determinants is presented relative to that of mock treatment (**B**) or control subline (**C**). * $p < 0.05$ (vs. mock or control).

We next examined if AR could regulate the expression of GABBR2, using a chromatin immunoprecipitation (ChIP) assay (Figure 2). A bioinformatics-driven search detected a potential binding site of AR in the promoter region of the *GABBR2* gene. DNA fragments derived from UMUC3 cells immunoprecipitated with an anti-AR antibody were amplified by PCR with a set of GABBR2 promoter-specific primers. The PCR product was visualized from those precipitated by the AR antibody, but not control precipitations, indicating that AR could interact with the GABBR2 promoter.

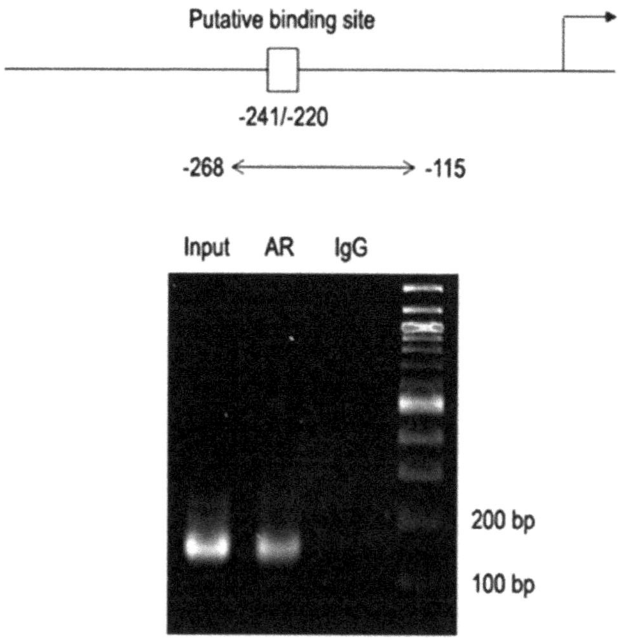

Figure 2. AR binding to the GABBR2 promoter in bladder cancer cells. The ChIP assay, using UMUC3 cell lysates immunoprecipitated with an anti-AR antibody (or IgG as a negative control). The DNA fragments were PCR amplified with a set of GABBR2 promoter-specific primers, and the PCR products were electrophoresed on 1% agarose gel. A fraction of the mixture of protein-DNA complex (i.e., 1% of total cross-linked, reserved chromatin prior to immunoprecipitation) was used as "input" DNA.

2.2. Role of GABBR2 in Cell Growth

Using its inhibitor and knockdown, we assessed the impact of GABBR2 on the growth of bladder cancer cells. Western blot showed that a selective GABA B receptor antagonist, CGP46381 (50% inhibitory concentration (IC50): 4.9 µM [25]), inhibited GABBR2 expression in UMUC3 cells (Figure 3A). Additionally, as expected, the level of GABBR2 expression in a UMUC3 subline stably expressing GABBR2-shRNA was substantially reduced, compared with control-shRNA-expressing cells (Figure 3B). Nonetheless, CGP46381 treatment in three AR-positive bladder cancer cell lines, as well as GABBR2 knockdown in UMUC3 cells, did not significantly change their proliferation (via MTT assay (Figure 4)) or migration (via wound-healing assay (Figure 5)).

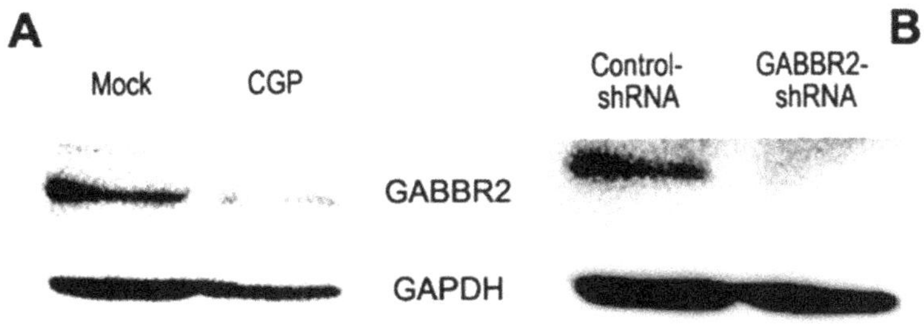

Figure 3. Effects of GABBR2 inhibitor or knockdown in bladder cancer cells. Western blotting of GABBR2 in UMUC3 cells cultured for 24 h with ethanol (mock) or CGP46381 (**A**) or UMUC3-control-shRNA vs. UMUC3-GABBR2-shRNA sublines (**B**).

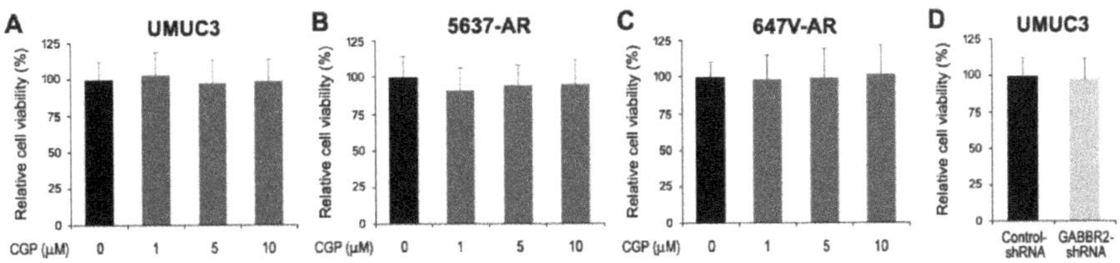

Figure 4. Effects of GABBR2 inhibitor or knockdown on the viability of bladder cancer cells. The MTT assay in UMUC3 (**A**), 5637-AR (**B**), or 647V-AR (**C**) cultured in the absence or presence of various doses of CGP46381 or UMUC3-control-shRNA vs. UMUC3-GABBR2-shRNA (**D**) for 48 h. Cell viability representing the mean (+SD) from a total of six determinants is presented relative to that of mock treatment or control subline.

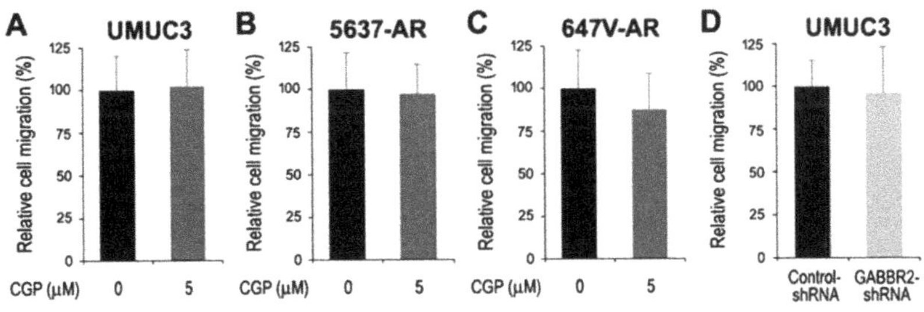

Figure 5. Effects of GABBR2 inhibitor or knockdown on the migration of bladder cancer cells. The wound-healing assay in UMUC3 (**A**), 5637-AR (**B**), or 647V-AR (**C**) without or with CGP46381 (5 μM) or UMUC3-control-shRNA vs. UMUC3-GABBR2-shRNA (**D**) cultured for 24 h after scratching. Cell migration determined by the rate of cells filling the wound area is presented relative to that of mock treatment or control subline. Each value represents the mean (+SD) from a total of six determinants.

2.3. Role of GABBR2 in CDDP Sensitivity

We then determined if GABBR2 could modulate sensitivity to CDDP treatment in bladder cancer cells. The MTT assay was first performed in UMUC3-control-shRNA and UMUC3-GABBR2-shRNA sublines treated with various doses of CDDP covering its pharmacological concentrations (i.e., 1.3–8.4 μM [26]). GABBR2-knockdown cells were significantly more sensitive to CDDP at 2–7.5 μM, and the IC50s were 4.8 μM in UMUC3-control-shRNA vs. 2.7 μM in UMUC3-GABBR2-shRNA (Figure 6).

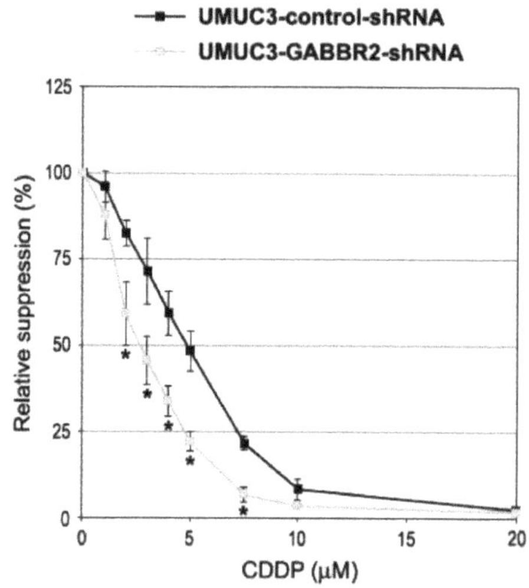

Figure 6. Effects of GABBR2 knockdown on CDDP cytotoxicity in bladder cancer cells. The MTT assay in UMUC3-control-shRNA vs. UMUC3-GABBR2-shRNA sublines cultured for 48 h in the presence of 0–20 μM CDDP. Cell viability representing the mean (±SD) from a total of six determinants is presented relative to that of each subline without CDDP treatment. * $p < 0.05$ (vs. control-shRNA).

We further assessed the effects of CGP46381 on CDDP cytotoxicity. In UMUC3 (Figure 7A), 5637-AR (Figure 7B), and 647V-AR (Figure 7C) cells, CGP46381 treatment enhanced the cytotoxic effects of CDDP over mock treatment. However, no significant effects of CGP46381 on CDDP suppression were seen in GABBR2-knockdown cells (Figure 7D).

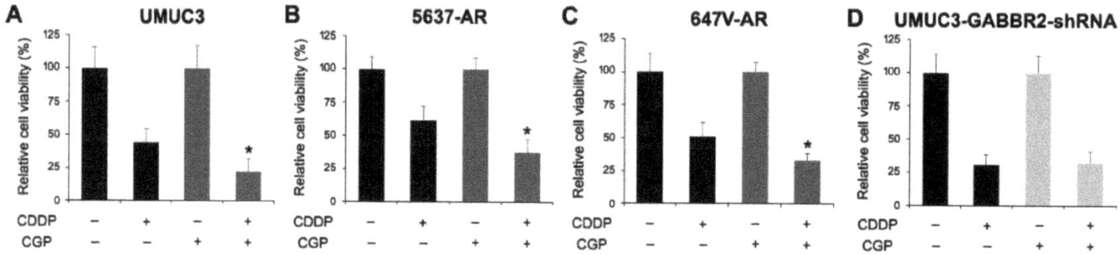

Figure 7. Effects of GABBR2 inhibitor on CDDP cytotoxicity in bladder cancer cells. The MTT assay in UMUC3 (**A**), 5637-AR (**B**), or 647V-AR (**C**) without or with CGP46381 (5 μM) or UMUC3-control-shRNA or UMUC3-GABBR2-shRNA (**D**) cultured for 48 h in the absence or presence of CDDP (5 μM). Cell viability representing the mean (+SD) from a total of six determinants is presented relative to that of mock or CGP46381 treatment (without vs. with CDDP) in each subline. * $p < 0.05$ [vs. CDDP(+)/CGP46381(−) cells].

2.4. Prognostic Value of GABBR2 in Bladder Cancer Patients

A publicly available database was searched for analyzing the prognostic significance of *GABBR2* expression in bladder cancer. Data obtained from GSE32894 [27] showed that the rate of overall survival was significantly lower in the entire cohort of patients with high *GABBR2* tumor (Figure 8A). When excluding those with noninvasive tumor who had been much less likely to die of bladder cancer, high *GABBR2* expression was still associated with a significantly higher mortality rate (Figure 8B).

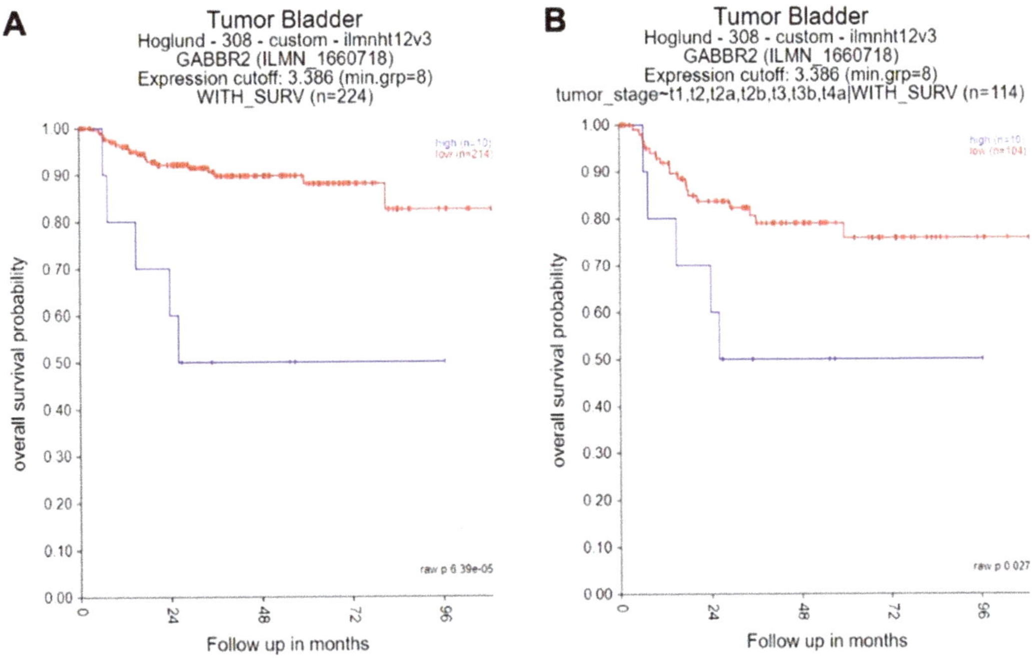

Figure 8. Kaplan–Meier curves for overall survival, according to the levels of *GABBR2* expression in the entire cohort of patients with bladder cancer (**A**) or those with invasive (stage T1 or higher) tumor (**B**).

3. Discussion

The efficacy of CDDP-based combined systemic chemotherapy in patients with urothelial cancer is often limited due to the initial failure or the acquisition of drug resistance during treatment, for which underlying molecular mechanisms remain not fully understood. Meanwhile, AR activation in bladder cancer cells has been associated with induction of resistance to CDDP [21–23,28–30] as well as other chemotherapeutic agents [31,32]. In the present study, we further investigated the functional role of GABBR2 in CDDP resistance in bladder cancer in relation to AR signaling.

GABA B receptors, including GABBR1 and GABBR2, are G protein-coupled receptor subunits. As GABA represents the main neurotransmitter in the brain, GABBR1 and/or GABBR2 have been implicated in the pathogenesis of various neurological or psychiatric disorders such as epilepsy, Huntington's disease, Alzheimer's disease, autism, and depression [33]. By contrast, little is known about the role of GABBR2 in neoplastic conditions. Limited and somehow conflicting data have demonstrated that GABA B receptor agonists either promote (e.g., breast [34], kidney [35], and prostate [36] cancers) or inhibit (e.g., lung cancer [37], cholangiocarcinoma [38,39]) the cell growth of nonurothelial malignancies. Particularly, in the prostate cancer study [36], an AR-negative cell line was tested, although the role of AR signaling in its progression and treatment has been extremely well established. In other studies, however, GABA B receptor agonists have been shown to have no significant effects on the cell proliferation and/or migration/invasion of colon [40] and liver [41] cancer lines as well as both AR-positive and AR-negative prostate cancer lines [42]. In a more recent study, a microRNA miR-31-3p was shown to suppress the growth of prostate cancer cells via directly downregulating GABBR2 expression similarly in AR-positive and AR-negative lines [43]. Thus, the involvement of GABBR2 in urothelial cancer progression or chemoresistance, as well as that in AR signaling, remains largely unknown.

As aforementioned, it remains unanswered how AR signaling modulates chemosensitivity. Based on the findings in DNA microarray analysis in control AR-positive vs. AR-knockdown bladder cancer sublines we had previously performed [24], we anticipated that GABBR2 functioned as a downstream target of AR. Herein, we demonstrated that AR expression or activity was positively correlated with GABBR2 expression. Specifically, androgen treatment and AR knockdown/antiandrogen treatment induced and reduced, respectively, the levels of GABBR2 expression in bladder cancer cells. Notably, a ChIP assay revealed an interaction of AR with GABBR2 at its promoter region, indicating the direct regulation of GABBR2 expression by AR. We then found that knockdown of GABBR2 via its shRNA or its inactivation via a GABA B receptor antagonist enhanced the cytotoxic effects of CDDP in AR-positive bladder cancer cells. Because the inhibitor failed to induce CDDP sensitivity in GABBR2-knockdown cells, its effect on CDDP cytotoxicity was likely mediated through GABBR2. However, GABBR2 did not appear to affect bladder cancer cell proliferation and migration in the absence of CDDP. These observations suggest that activation of AR induces CDDP resistance via directly upregulating the expression of GABBR2 in bladder cancer cells. Interestingly, in several nonurothelial cancer cells, cooperation of GABBR2 with EGFR-ERK1/2 signaling has been documented [35–38,44], while the ERK1/2 pathway has been linked to CDDP resistance [29,45]. Meanwhile, we have separately demonstrated that androgens activate the EGFR-ERK1/2 pathway in AR-positive bladder cancer cells [46].

GABBR2 expression has not been extensively studied in clinical samples. Nonetheless, the elevated expression of GABBR2 has been reported in several types of neoplasm such as thyroid adenoma and carcinoma [47] and non-small cell lung cancer [37]. In cholangiocarcinoma specimens, high GABBR2 expression was strongly associated with more aggressive nonpapillary histology yet small tumor size [39]. By analyzing a public database (i.e., The Cancer Genome Atlas (TCGA)), *GABBR2* overexpression was linked to lower T and N staging categories and better prognosis in patients with breast cancer [48]. Similarly, using a publicly available database [27], we here found that those with invasive bladder cancer showing high *GABBR2* expression had a significantly higher risk of overall mortality. This observation indeed supports our in vitro data indicating that GABBR2 could induce CDDP resistance in bladder cancer, although only a portion of patients in the database cohort may have undergone CDDP-based chemotherapy.

Again, in the present study, we have assessed if GABBR2 functions as a downstream effector of AR and, thereby, contributes to modulating sensitivity to CDDP. We first demonstrated that AR directly regulated the expression of GABBR2 via binding to its promoter in bladder cancer cells. AR inactivation, particularly via antiandrogen treatment, thus resulted in downregulation of GABBR2 expression. We then found that GABBR2 was involved in promoting CDDP resistance in bladder cancer cells. An antagonist for GABBR2 at the dose of IC50 thus significantly increased sensitivity to CDDP therapy. These findings suggest that the concurrent treatment with not only an antiandrogen, which has been widely used in patients with, for example, prostate cancer, but also a GABBR2 inhibitor, may considerably enhance the efficacy of CDDP-based chemotherapy, particularly in patients with AR-positive/GABBR2-positive bladder cancer. However, because an available inhibitor, CGP46381, is not entirely specific for GABBR2, the involvement of another GABA B receptor, GABBR1, in modulating CDDP sensitivity may need to be determined. Meanwhile, molecular subtypes of bladder cancer have been well established (e.g., TCGA cohort study [49]) and certain subtypes of muscle-invasive bladder cancer have been associated with chemoresistance [50]. It might, therefore, be of importance to determine which molecular subgroup of patients could generally benefit from such combination therapy.

4. Materials and Methods

4.1. Antibodies and Chemicals

We obtained anti-AR (441), anti-GABBR2 (H-10), and anti-GAPDH (6c5) antibodies from Santa Cruz Biotechnology (Dallas, TX, USA). DHT and HF, CGP46381, and CDDP were obtained from Sigma-Aldrich (St. Louis, MO, USA), Thermo Fisher (Waltham, MA, USA), and Santa Cruz Biotechnology (Dallas, TX, USA), respectively.

4.2. Cell Lines

A human urothelial carcinoma cell line, UMUC3, was originally obtained from the American Type Culture Collection and recently authenticated by the institutional core facility. Human urothelial cancer sublines stably expressing AR-shRNA (e.g., UMUC3-AR-shRNA) or control-shRNA (e.g., UMUC3-control-shRNA) [47], as well as human wild-type AR (e.g., 5637-AR [47], 647V-AR [51]), were established in our previous studies. Similarly, GABBR2-shRNA (sc-42463-V; Santa Cruz Biotechnology), a pool of concentrated, transduction-ready lentiviral particles containing three target-specific constructs encoding 19–25 nt shRNA, was stably expressed in UMUC3 cells. In addition, a CDDP-resistant subline (i.e., UMUC3-CR) and its control were previously established by long-term (i.e., >12 weeks), continuous, stepwise (i.e., 0.2–2.0 µM) treatment of CDDP [21]. These parental line and sublines were maintained in DMEM (Thermo Fisher) supplemented with 10% fetal bovine serum (FBS), penicillin (50 U/mL), and streptomycin (50 µg/mL) at 37 °C in a humidified atmosphere of 5% CO_2. The cells were then cultured in phenol red-free medium supplemented with 5% FBS (or 5% charcoal-stripped FBS for DHT/HF treatment) at least 24 h before actual assays.

4.3. Real-Time PCR

Total RNA isolated from cultured cells by TRIzol (Invitrogen, Waltham, MA, USA) was reverse transcribed using oligo (dT) primers and Omniscript reverse transcriptase (Qiagen, Germantown, MD, USA). Real-time PCR was then performed using iQ™ SYBR® Green Supermix (Bio-Rad, Hercules, CA, USA), as we described previously [21,24]. The following primer pairs were used: *GABBR2* (forward, 5′-TGGAGGCGTCTGTCCATCCGT-3′; reverse, 5′-GTCTTGCGTCAGCGTGCCCA-3′); and *GAPDH* (forward, 5′-AAGGTGAAGGTCGG AGTCAAC-3′; reverse, 5′-GGGGTCATTGATGGCAACAATA-3′).

4.4. Western Blot

Equal amounts of proteins (30 µg) extracted from cell extracts were subjected to electrophoresis with 10% sodium dodecyl sulfate-polyacrylamide gel, which was transferred to polyvinylidene difluoride membrane electronically. After blocking with 5% nonfat dry milk, the membrane was incubated with a primary antibody at 4 °C overnight, followed by 1 h incubation with an HRP-conjugated secondary antibody (Cell Signaling Technology, Danvers, MA, USA) at room temperature. Chemiluminescent signals were then generated by Clarity Western ECL Substrate and detected by ChemiDOC™ MP (Bio-Rad).

4.5. ChIP

We first performed a bioinformatic search for potential AR binding sites in the GABBR2 promoter (https://biogrid-lasagna.engr.uconn.edu/lasagna_search/ [52], accessed on 2 March 2022) (see Figure 2). ChIP assay was then performed, using a Magna ChIP kit (Sigma-Aldrich), according to the manufacturer's recommended protocol with minor modifications, as we described previously [30]. Briefly, cells were cross-linked with 1% formaldehyde, and the lysates were sonicated in nuclear buffer. Soluble chromatin was immunoprecipitated with an anti-AR antibody or normal mouse immunoglobulin G (IgG) (Santa Cruz Biochemistry) directly conjugated with magnetic protein A beads. Immunoprecipitated DNA was eluted and reverse cross-linked. DNA was then extracted and purified, using a spin filter column, and amplified by PCR, using the primers as follows: forward, 5′-ACGCCTTCGTGGAACATT-3′; reverse, 5′-ACAGCGATCTGGGAACC-3′. The PCR products electrophoresed on 1% agarose gel and stained with ethidium bromide were visualized, using Gel Doc XR + (Bio-Rad).

4.6. Cell Proliferation

The MTT assay was used to assess the cell viability. Cells (5×10^3/well) seeded in 96-well tissue culture plates were cultured for 48 h and then incubated with 0.5 mg/mL of MTT (3-(4,5-dimethylthiazol-2-yl)-2,5diphenyltetrazolium bromide; Sigma-Aldrich) for 3 h at 37 °C. MTT was dissolved by dimethyl sulfoxide, and the absorbance was measured at a wavelength of 570 nm with background subtraction at 630 nm. IC50 was calculated using the web-based tool (https://www.aatbio.com/tools/ic50-calculator, accessed on 7 August 2023).

4.7. Cell Migration

Scratch wound healing assay was adapted to evaluate the ability of cell migration. Cells at a density of ≥90% confluence in 6-well tissue culture plates were scratched manually with a sterile 200 µL plastic pipette tip. The wounded monolayers of the cells were allowed to heal in serum-free medium for 24 h, and the width of the wound area was monitored with an inverted microscope. The normalized cell-free area in photographed pictures (24 h/0 h) was quantitated using ImageJ software (National Institutes of Health, Bethesda, MD, USA).

4.8. Public Database Analysis

The R2 Genomics Analysis and Visualization Platform (https://hgserver1.amc.nl/cgi-bin/r2/main.cgi, accessed on 7 August 2023) was used to assess the prognostic role of *GABBR2* expression in patients with bladder cancer. Data were obtained from the Gene Expression Omnibus repository (GSE32894; https://www.ncbi.nlm.nih.gov/geo/query/acc.cgi?acc=GSE32894 [27], accessed on 7 August 2023).

4.9. Statistical Analysis

Student's *t*-test was used to compare numerical data. *p* values less than 0.05 were considered to be statistically significant.

5. Conclusions

We identified GABBR2 as a key downstream effector of AR in modulating CDDP sensitivity in bladder cancer. Our data suggest that not only concurrent antiandrogen therapy but also GABBR2 inhibitor treatment has the potential of being a means of chemosensitization, especially in patients with AR/GABBR2-positive bladder tumor. Nonetheless, further studies are warranted for elucidating molecular mechanisms responsible for AR/GABBR2-mediated chemoresistance.

Author Contributions: Conceptualization, H.M.; data acquisition, M.A.E.N., M.Y., Y.T., T.T. and G.J.; data analysis and interpretation, M.A.E.N., M.Y., Y.T., T.T., G.J. and H.M.; writing—original draft preparation, M.A.E.N. and M.Y.; writing—review and editing, Y.T., T.T. and H.M.; funding acquisition, H.M. All authors have read and agreed to the published version of the manuscript.

Funding: This research received no external funding.

Institutional Review Board Statement: Not applicable.

Informed Consent Statement: Not applicable.

Data Availability Statement: The data presented in this study are available on request from the corresponding author but are not publicly available (except public database analysis data—Section 2.4).

Conflicts of Interest: The authors declare no conflict of interest.

References

1. Torre, L.A.; Bray, F.; Siegel, R.L.; Ferlay, J.; Lortet-Tieulent, J.; Jemal, A. Global cancer statistics, 2012. *CA Cancer J. Clin.* **2015**, *65*, 87–108. [CrossRef] [PubMed]
2. Bray, F.; Ferlay, J.; Soerjomataram, I.; Siegel, R.L.; Torre, L.A.; Jemal, A. Global cancer statistics 2018: GLOBOCAN estimates of incidence and mortality worldwide for 36 cancers in 185 countries. *CA Cancer J. Clin.* **2018**, *68*, 397–424. [CrossRef] [PubMed]
3. SEER Cancer Stat Facts: Bladder Cancer. National Cancer Institute: Bethesda, MD, USA. Available online: http://seer.cancer.gov/statfacts/html/urinb.html (accessed on 29 August 2023).
4. Rouprêt, M.; Babjuk, M.; Compérat, E.; Zigeuner, R.; Sylvester, R.J.; Burger, M.; Cowan, N.C.; Gontero, P.; Van Rhijn, B.W.G.; Mostafid, A.H.; et al. European Association of Urology guidelines on upper urinary tract urothelial carcinoma: 2017 update. *Eur. Urol.* **2018**, *73*, 111–122. [CrossRef]
5. Hanna, K.S. Updates and novel treatments in urothelial carcinoma. *J. Oncol. Pharm. Pract.* **2019**, *25*, 648–656. [CrossRef]
6. Gust, K.M.; Resch, I.; D'Andrea, D.; Shariat, S.F. Update on systemic treatment of upper tract urothelial carcinoma: A narrative reviewe of the literature. *Transl. Androl. Urol.* **2021**, *10*, 4051–4061. [CrossRef]
7. Roviello, G.; Catalano, M.; Santi, R.; Santoni, M.; Galli, I.C.; Amorosi, A.; Polom, W.; De Giorgi, U.; Nesi, G. Neoadjuvant treatment in muscle-invasive bladder cancer: From the beginning to the latest developments. *Front. Oncol.* **2022**, *12*, 912699. [CrossRef]
8. Esteban-Villarrubia, J.; Torres-Jiménez, J.; Bueno-Bravo, C.; García-Mondaray, R.; Subiela, J.D.; Gajate, P. Current and future landscape of perioperative treatment for muscle-invasive bladder cancer. *Cancers* **2023**, *15*, 566. [CrossRef] [PubMed]
9. Hamid, A.R.A.H.; Ridwan, F.R.; Parikesit, D.; Widia, F.; Mochtar, C.A.; Umbas, R. Meta-analysis of neoadjuvant chemotherapy compared to radical cystectomy alone in improving overall survival of muscle-invasive bladder cancer patients. *BMC Urol.* **2020**, *20*, 158. [CrossRef]
10. Zargar, H.; Espiritu, P.N.; Fairey, A.S.; Mertens, L.S.; Dinney, C.P.; Mir, M.C.; Krabbe, L.M.; Cookson, M.S.; Jacobsen, N.E.; Gandhi, N.M.; et al. Multicenter assessment of neoadjuvant chemotherapy for muscle-invasive bladder cancer. *Eur. Urol.* **2015**, *67*, 241–249. [CrossRef] [PubMed]
11. Voskuilen, C.S.; Oo, H.Z.; Genitsch, V.; Smit, L.A.; Vidal, A.; Meneses, M.; Necchi, A.; Colecchia, M.; Xylinas, E.; Fontugne, J.; et al. Multicenter validation of histopathologic tumor regression grade after neoadjuvant chemotherapy in muscle-invasive bladder carcinoma. *Am. J. Surg. Pathol.* **2019**, *43*, 1600–1610. [CrossRef]
12. von der Maase, H.; Hansen, S.W.; Roberts, J.T.; Dogliotti, L.; Oliver, T.; Moore, M.J.; Bodrogi, I.; Albers, P.; Knuth, A.; Lippert, C.M.; et al. Gemcitabine and cisplatin versus methotrexate, vinblastine, doxorubicin, and cisplatin in advanced or metastatic bladder cancer: Results of a large, randomized, multinational, multicenter, phase III study. *J. Clin. Oncol.* **2023**, *41*, 3881–3890. [CrossRef]
13. Chang, F.L.; Lai, M.D. Various forms of mutant p53 confer sensitivity to cisplatin and doxorubicin in bladder cancer cells. *J. Urol.* **2001**, *166*, 304–310. [CrossRef]
14. Galluzzi, L.; Vitale, I.; Michels, J.; Brenner, C.; Szabadkai, G.; Harel-Bellan, A.; Castedo, M.; Kroemer, G. Systems biology of cisplatin resistance: Past, present and future. *Cell Death Dis.* **2014**, *5*, e1257. [CrossRef]
15. Sarin, N.; Engel, F.; Rothweiler, F.; Cinatl, J.; Michaelis, M.; Frötschl, R.; Frötschl, H.; Kalayda, G.V. Key players of cisplatin resistance: Towards a systems pharmacology approach. *Int. J. Mol. Sci.* **2018**, *19*, 767. [CrossRef]
16. Cai, Z.; Zhang, F.; Chen, W.; Zhang, J.; Li, H. miRNAs: A promising target in the chemoresistance of bladder cancer. *OncoTargets Ther.* **2019**, *12*, 11805–11816. [CrossRef] [PubMed]
17. Tchounwou, P.B.; Dasari, S.; Noubissi, F.K.; Ray, P.; Kumar, S. Advances in our understanding of the molecular mechanisms of action of cisplatin in cancer therapy. *J. Exp. Pharmacol.* **2021**, *13*, 303–328. [CrossRef] [PubMed]
18. Long, G.; Ma, S.; Shi, R.; Sun, Y.; Hu, Z.; Chen, K. Circular RNAs and drug resistance in genitourinary cancers: A literature review. *Cancers* **2022**, *14*, 866. [CrossRef] [PubMed]
19. Inoue, S.; Mizushima, T.; Miyamoto, H. Role of the androgen receptor in urothelial cancer. *Mol. Cell. Endocrinol.* **2018**, *465*, 73–81. [CrossRef] [PubMed]
20. Ide, H.; Miyamoto, H. Sex hormone receptor signaling in bladder cancer: A potential target for enhancing the efficacy of conventional non-surgical therapy. *Cells* **2021**, *10*, 1169. [CrossRef]
21. Kashiwagi, E.; Ide, H.; Inoue, S.; Kawahara, T.; Zheng, Y.; Reis, L.O.; Baras, A.S.; Miyamoto, H. Androgen receptor activity modulates responses to cisplatin treatment in bladder cancer. *Oncotarget* **2016**, *7*, 49169–49179. [CrossRef] [PubMed]
22. Sekino, Y.; Sakamoto, N.; Ishikawa, A.; Honma, R.; Shigematsu, Y.; Hayashi, T.; Sentani, K.; Oue, N.; Teishima, J.; Matsubara, A.; et al. Transcribed ultraconserved region Uc.63+ promotes resistance to cisplatin through regulation of androgen receptor signaling in bladder cancer. *Oncol. Rep.* **2019**, *41*, 3111–3118. [CrossRef] [PubMed]
23. Tyagi, A.; Chandrasekaran, B.; Kolluru, V.; Rai, S.; Jordan, A.C.; Houda, A.; Messer, J.; Ankem, M.; Damodaran, C.; Haddad, A. Combination of androgen receptor inhibitor and cisplatin, an effective treatment strategy for urothelial carcinoma of the bladder. *Urol. Oncol.* **2019**, *37*, 492–502. [CrossRef] [PubMed]
24. Mizushima, T.; Jiang, G.; Kawahara, T.; Li, P.; Han, B.; Inoue, S.; Ide, H.; Kato, I.; Jalalizadeh, M.; Miyagi, E.; et al. Androgen receptor signaling reduces the efficacy of bacillus Calmette-Guèrin therapy for bladder cancer via modulating Rab27b-induced exocytosis. *Mol. Cancer Ther.* **2020**, *19*, 1930–1942. [CrossRef]

25. Olpe, H.R.; Steinmann, M.W.; Ferrat, T.; Pozza, M.F.; Greiner, K.; Brugger, F.; Froestl, W.; Mickel, S.J.; Bittiger, H. The actions of orally active GABA$_B$ receptor antagonists on GABAergic transmission in vivo and in vitro. *Eur. J. Pharmacol.* **1993**, *233*, 179–186. [CrossRef]
26. Kelsen, D.P.; Alcock, N.; Young, C.W. Cisplatin nephrotoxicity. Correlation with plasma platinum concentrations. *Am. J. Clin. Oncol.* **1985**, *8*, 77–80. [CrossRef]
27. Sjödahl, G.; Lauss, M.; Lövgren, K.; Chebil, G.; Gudjonsson, S.; Veerla, S.; Patschan, O.; Aine, M.; Fernö, M.; Ringnér, M.; et al. A molecular taxonomy for urothelial carcinoma. *Clin. Cancer Res.* **2012**, *18*, 3377–3386. [CrossRef] [PubMed]
28. Ide, H.; Goto, T.; Teramoto, Y.; Mizushima, T.; Jiang, G.; Nagata, Y.; Inoue, S.; Baras, A.S.; Kashiwagi, E.; Miyamoto, H. FOXO1 inactivation induces cisplatin resistance in bladder cancer. *Cancer Sci.* **2020**, *111*, 3397–3400. [CrossRef]
29. Jiang, G.; Teramoto, Y.; Goto, T.; Mizushima, T.; Inoue, S.; Ide, H.; Nagata, Y.; Kashiwagi, E.; Baras, A.S.; Netto, G.J.; et al. Identification of BXDC2 as a key downstream effector of the androgen receptor in modulating cisplatin sensitivity in bladder cancer. *Cancers* **2021**, *13*, 975. [CrossRef]
30. Teramoto, Y.; Jiang, G.; Goto, T.; Mizushima, T.; Nagata, Y.; Netto, G.J.; Miyamoto, H. Androgen receptor signaling induces cisplatin resistance via down-regulating GULP1 expression in bladder cancer. *Int. J. Mol. Sci.* **2021**, *22*, 10030. [CrossRef]
31. Shiota, M.; Takeuchi, A.; Yokomizo, A.; Kashiwagi, E.; Tatsugami, K.; Kuroiwa, K.; Naito, S. Androgen receptor signaling regulates cell growth and vulnerability to doxorubicin in bladder cancer cells. *J. Urol.* **2012**, *188*, 276–286. [CrossRef] [PubMed]
32. Kameyama, K.; Horie, K.; Mizutani, K.; Kato, T.; Fujita, Y.; Kawakami, K.; Kojima, T.; Miyazaki, T.; Deguchi, T.; Ito, M. Enzalutamide inhibits proliferation of gemcitabine-resistant bladder cancer cells with increased androgen receptor expression. *Int. J. Oncol.* **2017**, *50*, 75–84. [CrossRef]
33. Shaye, H.; Stauch, B.; Gati, C.; Cherezov, V. Molecular mechanisms of metabotropic GABA$_B$ receptor function. *Sci. Adv.* **2021**, *7*, eabg3362. [CrossRef] [PubMed]
34. Zhang, D.; Li, X.; Yao, Z.; Wei, C.; Ning, N.; Li, J. GABAergic signaling facilitates breast cancer metastasis by promoting ERK1/2-dependent phosphorylation. *Cancer Lett.* **2014**, *34*, 100–108. [CrossRef] [PubMed]
35. Inamoto, T.; Azuma, H.; Sakamoto, T.; Kiyama, S.; Ubai, T.; Kotake, Y.; Watanabe, M.; Katsuoka, Y. Invasive ability of human renal cell carcinoma cell line Caki-2 is accelerated by gamma-aminobutyric acid, via sustained activation of ERK1/2 inducible matrix metalloproteinases. *Cancer Investig.* **2007**, *25*, 574–583. [CrossRef]
36. Xia, S.; He, C.; Zhu, Y.; Wang, S.; Li, H.; Zhang, Z.; Jiang, X.; Liu, J. GABA$_B$R-induced EGFR transactivation promotes migration of human prostate cancer cells. *Mol. Pharmacol.* **2017**, *92*, 265–277. [CrossRef] [PubMed]
37. Zhang, X.; Zhang, R.; Zheng, Y.; Shen, J.; Xiao, D.; Li, J.; Shi, X.; Huang, L.; Tang, H.; Liu, J.; et al. Expression of gamma-aminobutyric acid receptors on neoplastic growth and prediction of prognosis in non-small cell lung cancer. *J. Transl. Med.* **2013**, *11*, 102. [CrossRef]
38. Fava, G.; Marucci, L.; Glaser, S.; Francis, H.; De Morrow, S.; Benedetti, A.; Alvaro, D.; Venter, J.; Meininger, C.; Patel, T.; et al. γ-Aminobutyric acid inhibits cholangiocarcinoma growth by cyclic AMP-dependent regulation of the protein kinase A/extracellular signal-regulated kinase 1/2 pathway. *Cancer Res.* **2005**, *65*, 11437–11446. [CrossRef]
39. Saengboonmee, C.; Sorin, S.; Sangkhamanon, S.; Chomphoo, S.; Indramanee, S.; Seubwai, W.; Thithuan, K.; Chiu, C.F.; Okada, S.; Gingras, M.C.; et al. γ-aminobutyric acid B2 receptor: A potential therapeutic target for cholangiocarcinoma in patients with diabetes mellitus. *World J. Gastroenterol.* **2023**, *29*, 4416–4432. [CrossRef] [PubMed]
40. Joseph, J.; Niggemann, B.; Zaenker, K.S.; Entschladen, F. The neurotransmitter gamma-aminobutyric acid is an inhibitory regulator for the migration of SW 480 colon carcinoma cells. *Cancer Res.* **2002**, *62*, 6467–6469.
41. Lodewyks, C.; Rodriguez, J.; Yan, J.; Lerner, B.; Lipschitz, J.; Nfon, C.; Rempel, J.D.; Uhanova, J.; Minuk, G.Y. GABA-B receptor activation inhibits the in vitro migration of malignant hepatocytes. *Can. J. Physiol. Pharmacol.* **2011**, *89*, 393–400. [CrossRef]
42. Abdul, M.; Mccray, S.D.; Hoosein, N.M. Expression of gamma-aminobutyric acid receptor (subtype A) in prostate cancer. *Acta Oncol.* **2008**, *47*, 1546–1550. [CrossRef] [PubMed]
43. Choi, S.; Lee, S.; Han, Y.H.; Choi, J.; Kim, I.; Lee, J.; An, H.J. miR-31-3p functions as a tumor suppressor by directly targeting GABBR2 in prostate cancer. *Front. Oncol.* **2022**, *12*, 945057. [CrossRef] [PubMed]
44. Niu, X.; Liu, F.; Zhou, Y.; Zhou, Z.; Zhou, D.; Wang, T.; Li, Z.; Ye, X.; Yu, Y.; Weng, X.; et al. Genome-wide DNA methylation analysis reveals *GABBR2* as a novel epigenetic target for EGFR 19 deletion lung adenocarcinoma with induction erlotinib treatment. *Clin. Cancer Res.* **2017**, *23*, 5003–5014. [CrossRef] [PubMed]
45. Achkar, I.W.; Abdulrahman, N.; Al-Sulaiti, H.; Joseph, J.M.; Uddin, S.; Mraiche, F. Cisplatin based therapy: The role of the mitogen activated protein kinase signaling pathway. *J. Transl. Med.* **2018**, *16*, 96. [CrossRef] [PubMed]
46. Zheng, Y.; Izumi, K.; Yao, J.L.; Miyamoto, H. Dihydrotestosterone upregulates the expression of epidermal growth factor receptor and ERBB2 in androgen receptor-positive bladder cancer cells. *Endocr. Relat. Cancer* **2011**, *18*, 451–464. [CrossRef]
47. Roberts, S.S.; Mendonça, M.C.; Jensen, K.; Francis, G.L.; Vasko, V. GABA receptor expression in benign and malignant thyroid tumors. *Pathol. Oncol. Res.* **2009**, *15*, 645–650. [CrossRef]
48. Liu, S.; Song, A.; Zhou, X.; Huo, Z.; Yao, S.; Yang, B.; Liu, Y.; Wang, Y. ceRNA network development and tumour-infiltrating immune cell analysis of metastatic breast cancer to bone. *J. Bone Oncol.* **2020**, *24*, 100304. [CrossRef] [PubMed]
49. Robertson, A.G.; Kim, J.; Al-Ahmadie, H.; Bellmunt, J.; Guo, G.; Cherniack, A.D.; Hinoue, T.; Laird, P.W.; Hoadley, K.A.; Akbani, R.; et al. Comprehensive molecular characterization of muscle-invasive bladder cancer. *Cell* **2017**, *171*, 540–556.e25. [CrossRef]

50. Choi, W.; Porten, S.; Kim, S.; Willis, D.; Plimack, E.R.; Hoffman-Censits, J.; Roth, B.; Cheng, T.; Tran, M.; Lee, I.L.; et al. Identification of distinct basal and luminal subtypes of muscle-invasive bladder cancer with different sensitivities to frontline chemotherapy. *Cancer Cell* **2014**, *25*, 152–165. [CrossRef]
51. Li, Y.; Ishiguro, H.; Kawahara, T.; Miyamoto, Y.; Izumi, K.; Miyamoto, H. GATA3 in the urinary bladder: Suppression of neoplastic transformation and down-regulation by androgens. *Am. J. Cancer Res.* **2014**, *4*, 461–473.
52. Lee, C.; Huang, C.H. LASAGNA-Search 2.0: Integrated transcription factor binding site search and visualization in a browser. *Bioinformatics* **2014**, *30*, 1923–1925. [CrossRef] [PubMed]

International Journal of
Molecular Sciences

MDPI

Article

Interleukin-1β/Interleukin (IL)-1-Receptor-Antagonist (IL1-RA) Axis in Invasive Bladder Cancer—An Exploratory Analysis of Clinical and Tumor Biological Significance

Marko Vukovic [1,2], Jorge M. Chamlati [1], Jörg Hennenlotter [1], Tilman Todenhöfer [1], Thomas Lütfrenk [1], Sebastian Jersinovic [1], Igor Tsaur [1], Arnulf Stenzl [1] and Steffen Rausch [1,*]

1 Department of Urology, Eberhard-Karls-University, 72074 Tuebingen, Germany; s.jersinovic@gmail.com (S.J.)
2 Department of Urology, Clinical Center of Montenegro, University of Montenegro, 81000 Podgorica, Montenegro
* Correspondence: steffen.rausch@med.uni-tuebingen.de

Abstract: Previous data indicate a role of IL-1 and IL-1RA imbalance in bladder carcinoma (BC); the inhibition of IL-1 signaling might be considered a treatment option. Objective: To assess expression patterns and the prognostic role of IL-1β and IL-1RA in invasive BC and to evaluate their interaction with AKT signaling and proliferation. The study included two independent cohorts of n = 92 and n = 102 patients who underwent a radical cystectomy for BC. Specimen from BC and benign urothelium (n = 22 and n = 39) were processed to a tissue microarray and immunohistochemically stained for IL-1β, IL-1RA, AKT, and Ki-67. Expression scores were correlated to clinical variables and Ki-67 and AKT expression. An association with outcome was assessed using Wilcoxon Kruskal–Wallis tests, Chi-square tests or linear regression, dependent on the variable's category. Kaplan–Meier and Cox proportional hazard analyses were used to estimate recurrence-free (RFS), cancer-specific (CSS) and overall survival (OS). Both IL-1β and IL-1RA were significantly overexpressed in invasive BC compared to benign urothelium in both cohorts ($p < 0.005$). IL-1β was associated with vascular invasion (210 vs. 183, $p < 0.02$), lymphatic invasion (210 vs. 180, <0.05) and G3 cancer (192 vs. 188, <0.04). The survival analysis revealed favorable RFS, CSS, and OS in the case of high IL-1β expression ($p < 0.02$, <0.03, and <0.006, respectively). Multivariate analyses revealed an independent impact of (low) IL1β expression on RFS, CSS, and OS. The IL-1β and IL-1β/IL-1RA ratios were positively correlated to the AKT expression ($p < 0.05$ and <0.01, respectively). Additionally, the high expression of Ki-67 (>15%) correlated with higher levels of IL-1β ($p = 0.01$). The overexpression of IL-1β and IL-1RA is frequently found in BC, with a prognostic significance observed for the IL-1β protein expression. The observed link between the IL-1β/IL-1RA axis and AKT signaling may indicate possible autophagy activation processes besides the known tumor-promoting effects of AKT.

Keywords: urothelial cancer; metastasis; prognosis; biomarker; pathway; autophagy

1. Introduction

Bladder cancer (BC) is a potentially life-threatening malignancy and is considered as one of the most expensive tumors in terms of treatment and medical care [1]. Several metabolic pathways are involved in BC tumorigenesis, representing potential targets for therapy [2]. Recent advances in the molecular understanding of urothelial cancer (UC) have led to the identification of prognostic biomarkers displaying an interplay between immunity and autophagy [3,4]. While Ki-67 expression is a well-established prognostic proliferation biomarker in BC [5], Beclin-1 and Protein kinase B (AKT)-mediated pathways have been recognized as important autophagy mediators in the development and progression of UC [6,7]). Recently proposed biomarkers, with diagnostic and predictive relevance in BC, are the interleukin–1 beta (IL-1β) and IL-1 receptor antagonist (IL-1RA) [8]. Several studies reported that IL-1RA induced a restoration of the autophagy process and catabolism reduction in inflammation-altered human cells, while IL-1β acted more like autophagy inhibitor, especially in cancer cells [9–11]. Despite the assumed value of both the IL-1β and IL-1RA status, only little is known about their (combined) prognostic value and on their correlation to cancer proliferation (Ki-67) and autophagy markers such as AKT. Therefore, the aim of the present study was to assess the expression patterns and prognostic role of the IL-1β/IL-1RA axis in invasive BC, as well as their correlation with Ki-67 and AKT in order to gain a further insight into their molecular relevance in BC. We hypothesize its role based on the high expression level of IL-1β within more invasive or aggressive cancer, as well as its association with survival. We further addressed the correlation of the IL axis to autophagy and proliferative activity, indicating a potential anti-tumorigenic role.

2. Results

2.1. Patients' Characteristics

In total, 194 subjects who underwent a radical cystectomy for invasive BC were retrospectively enrolled for this study. Patients were selected in two cohorts: 102 in a discovery and 92 patients in confirmatory group. The first group had a median follow-up of 44.5 months, and only within this cohort was the correlation with AKT examined. The correlation with Ki-67 was examined within the second cohort. Both groups were comparable in terms of age, gender, pathological stage, grade, and metastatic disease status. The baseline characteristics for both cohorts are shown in Table 1.

Table 1. Patients' characteristics.

Patient Characteristics	Group I (n = 102)		Group II (n = 92)		*p*
	N	%	n	%	
Gender					0.56
-Female	24	23.53	25	27.18	
-Male	78	76.47	67	72.82	
Age	Median 69	Range (59–74)	Median 65	Range (55–72)	0.19
Pathological stage					0.55
-T1	0	0.00	9	9.78	
-T2	31	30.39	21	22.82	
-T3	47	46.08	42	45.66	
-T4	24	23.53	18	19.57	
-Only CIS	0	0.00	2	2.17	
Lymph node status					0.50
-N0	58	56.86	57	61.95	
-N+	44	43.14	35	38.05	
Distant metastasis					0.59
-M0	89	87.23	82	89.13	
-M1	9	8.83	10	10.87	
-Mx	4	3.93	0	0.00	
Pathological grade					0.94
-G1	0	0.00	1	1.08	
-G2	24	23.53	20	21.74	
-G3	78	76.47	71	77.18	
Resection margin status					0.74
-R0	84	82.35	74	80.43	
-R1	15	14.70	14	15.22	
-R2	3	2.95	2	2.18	
-Rx	0	0.00	2	2.17	

Table 1. *Cont.*

Patient Characteristics	Group I (n = 102)		Group II (n = 92)		*p*
	N	%	n	%	
Carcinoma in situ (CIS)					0.57
-No	68	66.66	61	66.31	
-Yes	29	28.44	31	33.69	
-CISx	5	4.90	0	0.00	

2.2. Expression Characteristics of IL-1β and IL1RA

The mean cytoplasmic expression of IL-1β and IL-1RA in BC and surrounding benign urothelium tissue was 185 and 190 vs. 120 and 110 (p = 0.0001 and 0.004), respectively. Figure 1A–I shows representative histological images of IL-1β and IL-1RA expression in the histopathological benign and malignant urothelium. Although heterogeneity was observed with regard to the cytoplasmatic distribution and intensity of expression in BC, the staining for both IL-1β and IL-1RA in benign controls was predominantly of a low intensity. Figure 2A,B Illustrate differences in the intensity of the IL-1 expression pattern between normal and malignant tissue. Only a few benign urothelial tissues showed high expression levels of both IL-1β (H-score: 240) and IL-1RA (H-score: 280). In summary, both Il-1β as well as IL-1RA were significantly overexpressed in invasive UC compared to benign urothelium (both $p < 0.005$). These findings could be reproduced in the confirmatory cohort (all $p < 0.05$).

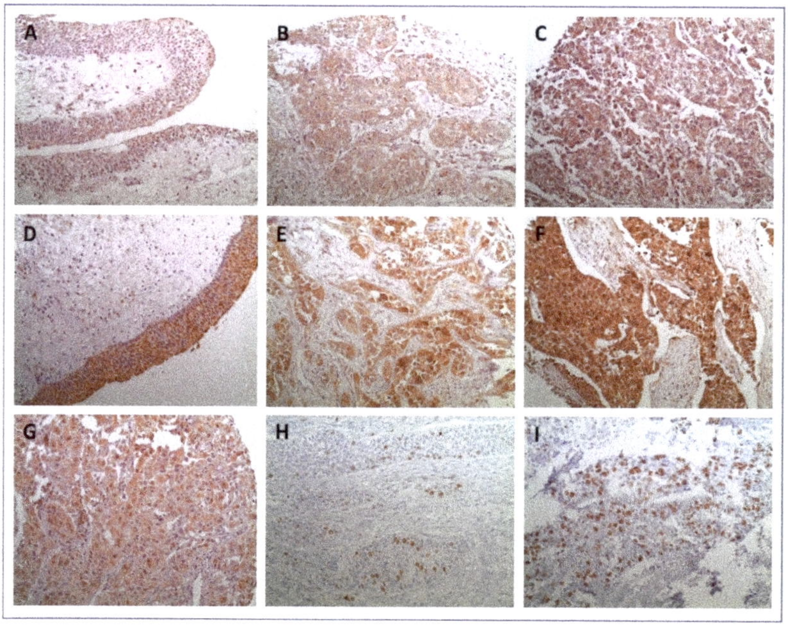

Figure 1. (**A–I**) Representative staining for immunohistochemistry of IL-1β/IL-1RA, AKT, and Ki-67 signaling pathway markers in bladder cancer and normal tissues. (**A–C**) Cytoplasmic IL-1β immunostaining in bladder tissue: (**A**)—staining in benign tissue; (**B,C**)—low and high staining in cancer tissue. (**D–F**) Cytoplasmic IL-1RA immunostaining in bladder tissue: (**D**)—staining in benign tissue; (**E,F**)—low and high staining in cancer tissue. (**G**) Cytoplasmic and partial nuclear staining of AKT in cancer tissue. (**H,I**) Low and strong (≥15%) nuclear staining of Ki-67 in cancer tissue. Magnification: 200-fold.

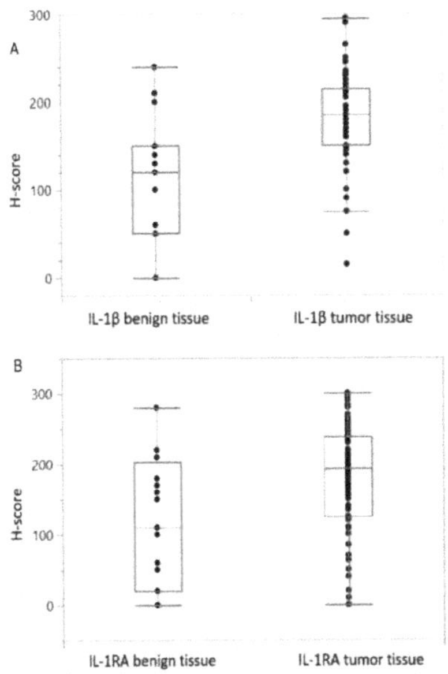

Figure 2. A–B. Expression patterns of IL-1β and IL-1RA. (**A**), Intensity of IL-1β cells expression and in (**B**), Intensity of IL-1RA cells expression in benign and malignant tissue; $p = 0.0001$ and $p = 0.004$, Wilcoxon Kruskal-Wallis tests).

2.3. Correlation of IL-1 Axis with Histopathologic Parameters

Table 2 summarizes the correlation of IL-1β and IL-1RA expression with histopathologic variables in a univariate analysis. Expression patterns were predictive of a pathological stage (≥T2) and nodal or distant metastasis (N+, M+). A strong IL-1β expression was observed in patients with vascular (210 vs. 183, $p < 0.02$) and lymphatic invasion (210 vs. 180, <0.05), as well as in G3 UC (193 vs. 188, <0.04). The expression of IL-1RA did not correlate with tumor stage, lymphovascular invasion, N+, or M+ status. Nevertheless, the significant association of the low expression of IL-1RA and high-grade UC ($p = 0.002$) was determined, which led to a positive correlation between the IL-1 beta/IL-1RA ratio and tumor grade ($p = 0.002$). Furthermore, G3 UC showed a lower median IL1-RA expression compared to G2 (185 vs. 250, <0.003), and the IL-1 beta/IL-1RA ratio was consequently significantly higher in G3 UC (1.10 vs. 0.66, $p = 0.001$).

Table 2. Correlations of IL-1β and IL-1RA expressions and IL-1 ratio with histopathological variables.

Histopathology Variables	IL-1β Expression (H-Score)	IL-1RA Expression (H-Score)	IL1-β/IL-1RA Ratio (Cut-Off 0.7)
Tumor grade	Medians		
G2	188	**250** ***	0.7
G3	**193** *	185	**1.1** ***
Tumor stadium pT			
T2	205	178	1.0
T3	190	200	1.0
T4	190	200	0.9

Table 2. *Cont.*

Histopathology Variables	IL-1β Expression (H-Score)	IL-1RA Expression (H-Score)	IL1-β/IL-1RA Ratio (Cut-Off 0.7)
Lymphatic invasion			
No	180	194	1.0
yes	**210 ***	200	1.0
Vascular invasion			
no	**183**	190	1.0
yes	**210 ***	203	1.0
Nodal invasion			
N0	185	194	1.0
N1	185	200	0.9
N2	210	205	1.1
N3	198	175	1.0
Metastatic disease			
M0	185	194	1.0
M1	210	200	1.0

Asterisk depicts statistically significant correlation between expression pattern of IL-1 axis and corresponding histological variable (* ≤ 0.05; *** ≤ 0.005).

2.4. Combined Scores and Clinical Data

Clinical parameters were compared with combined biomarker scores 'Ki-67+ and IL-1β+' and 'Ki-67+ or IL-1β+' as well as and/or IL-1RA, respectively. Statistical significance could be demonstrated in 'Ki-67+ and IL-1β+' for vascular invasion and in 'Ki-67+ or IL-1β+' and (higher) tumor grade (Table 3).

Table 3. Correlation of clinicopathological characteristics and combined biomarker scores in patients with urothelial carcinoma; *p*-values from chi-square tests. Ki-67+ describes expression level of >15%.

Variable	Ki-67+ and IL-1β+	Ki-67+ (15%) or IL-1β+	Ki-67+ and IL1RA+	Ki-67+ (15%) or IL-1RA+
Pathological stage >T2 vs. T2	0.61	0.67	0.81	0.06
Nodal status N+ vs. N0	0.97	0.25	0.39	0.54
Metastasis status M+ vs. M0	0.28	0.40	0.46	0.56
Grade ≥G3 vs. <G3	1.00	**0.02**	0.52	0.57
Lymphatic invasion L+ vs. L0	0.93	0.51	0.30	0.70
Vascular invasion V+ vs. V0	0.02	0.92	0.33	0.90

2.5. Correlation of IL-1 Axis with Ki-67 and AKT Expression

Regarding potential associations with proliferation or autophagy, IL-1β and IL-1 beta/IL-1RA ratio (cut-off 0.7) were associated with a higher AKT expression in UC cells ($p = 0.04$ and $p = 0.009$, respectively, Figure 3A,B). Moreover, high expression scores of Ki-67 (>15%) correlated with a higher expression of IL-1β ($p = 0.01$, Figure 3C). Although no statistical significance was observed for IL-1RA, there was a positive trend towards a correlation between high Ki-67 scores and a IL-1 beta/IL-1RA ratio ($p = 0.08$) (Figure 3D).

Moreover, a contingency analysis confirmed the significant association of combined expressions of 'Ki-67 and IL-1β', as well as a Ki-67/IL-1 ratio with a high tumor grade ($p = 0.02$ and $p = 0.01$, respectively—not shown). Postulated relationship between IL-1β/IL-1RA axis and autophagy signaling pathway in invasive bladder cancer has been illustrated in Figure 4.

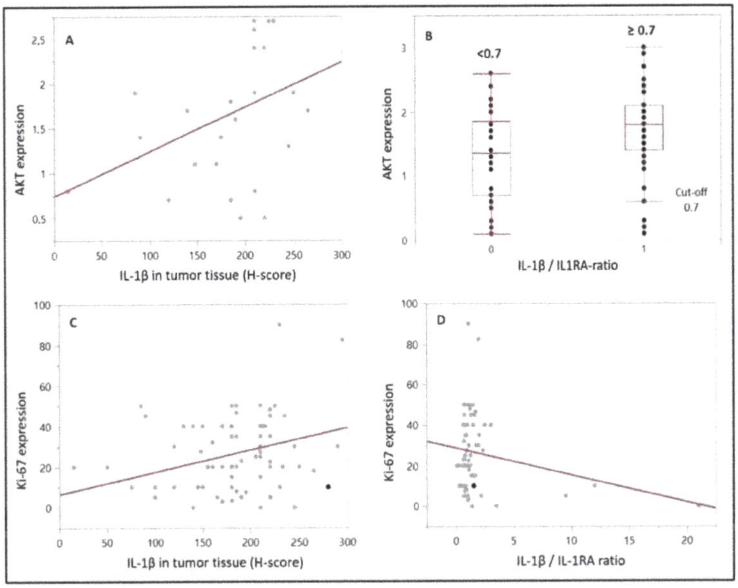

Figure 3. (**A–D**). (**A**), Correlation between expression patterns of IL-1β and AKT in tumor tissue ($p = 0.04$, r^2 corrected 0.14); (**B**), Positive correlation between AKT expression pattern and IL-1 ratio, for the determined cut-off value (0—under 0.7 cut-off; 1—over or = 0.7 cut-off); (**C**,**D**), Positive correlation between IL-1β expression and Ki-67 expression in tumor tissue ($p = 0.011$, $r^2 = 0.095$), as well as positive correlation trend of Ki-67 high expression towards IL-1 beta/IL-1RA ratio ($p = 0.08$).

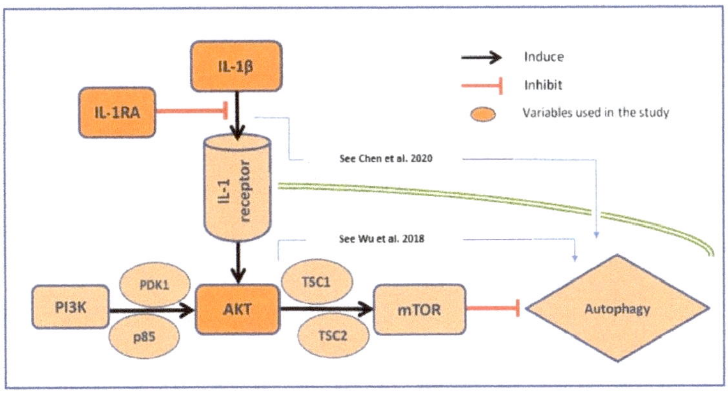

Figure 4. Postulated relationship between IL-1β/IL-1RA axis and autophagy signaling pathway in invasive bladder cancer. IL-interleukin, PIK3/AKT/mTOR—Phosphoinositidylinositol-3-Kinase/Proteinkinase B/mammalian Target of Rapamycin signaling pathway, PDK phosphoinositide-dependent protein kinase-1, TSC1/TSC2—Tuberous Sclerosis Complex 1 and 2, and p85—protein, the regulatory subunit of PI3K [12,13].

2.6. Survival Analysis

Figure 5 illustrates Kaplan–Meyer estimates for RFS, CSS, and OS in dependence of IL-1β expression levels in the discovery cohort. The survival analysis revealed an improved RFS, CSS as well as OS in case of a IL-1β high expression ($p < 0.02$, <0.03 and <0.006, respectively). No significant correlation could be observed with regard to the IL-1RA expression or IL-1β/IL1RA ratio and any of the survival outcomes.

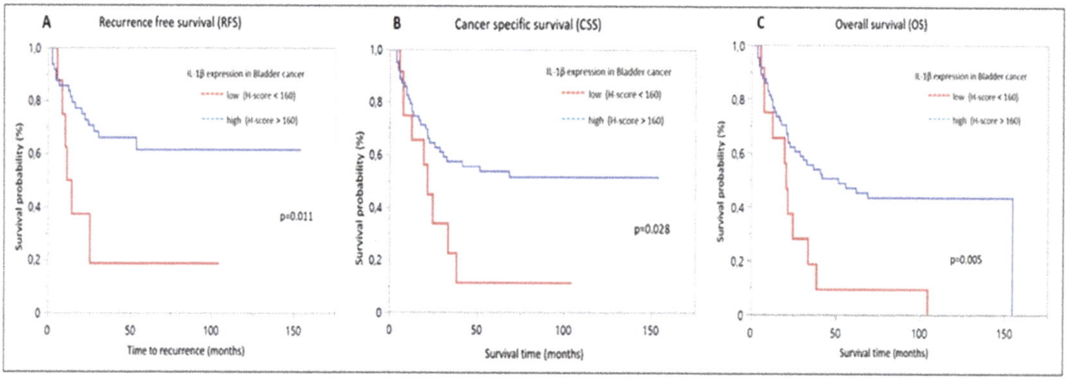

months	0	5	10	14	25	104	123
<160	8	8	6	4	2	1	
>160	49	43	41	39	32	10	6

0	12	24	38	104	154
12	12	4	2	1	
64	51	37	32	10	1

0	5	12	20	38	104	154
12	12	8	6	2	1	
65	60	51	43	32	10	1

Figure 5. A–C: Kaplan-Meier diagrams for IL-1β expression: (**A**). recurrence-free survival, (**B**). cancer specific survival and (**C**). overall survival; red = expression <160, log rank test, below numbers at risk.

2.7. Uni- and Multivariate Cox Proportional Hazard Analysis

Results are illustrated in Table 4. In the multivariate analysis, IL1β was confirmed as an independent predictor of RFS. All investigated parameters including age, T, N+, M, and G, as well as IL1-β, were significantly associated with CSS, and the multivariate analyses revealed an independent impact of IL1-β on CSS. Moreover, IL1-β was significantly associated with OS in univariable and multivariate analyses.

Table 4. Univariate and multivariate Cox regression analyses for recurrence-free survival (RFS), cancer-specific survival (CSS), and overall survival (OS) in the first cohort.

Variable	Univariate Analysis			Multivariate Analysis		
Recurrence (RFS)	*p*	**HR**	**95% CI**	*p*	**HR**	**95% CI**
Pathological stage >T2 vs. T2	**0.0099**	3.25	1.33–7.95	0.1339	2.14	0.79–5.80
Nodal status N+ vs. N0	0.0879	1.88	0.91–3.88		-	
Metastasis status M+ vs. M0		-			-	
Grade ≥G3 vs. <G3	0.0531	1.30	0.60–2.83		-	
IL-1β Low vs. high tumor expression (cut-off 160)	**0.0172**	3.16	1.23–8.16	**0.0265**	2.97	1.14–7.77
Age, year	0.2397	1.02	0.99–1.06		-	

Table 4. *Cont.*

Variable	Univariate Analysis			Multivariate Analysis		
Recurrence (RFS)	*p*	**HR**	**95% CI**	*p*	**HR**	**95% CI**
Variable	Univariate analysis			Multivariate analysis		
CSS	*p*	HR	95% CI	*p*	HR	95% CI
Pathological stage >T2 vs. T2	**0.0107**	2.53	1.24–5.17	0.3407	1.51	0.65–3.49
Nodal status N+ vs. N0	**0.0309**	1.96	1.06–3.62	0.3170	1.48	0.69–3.17
Metastasis status M+ vs. M0	**0.0015**	3.56	1.63–7.79	0.0504	2.58	1.00–6.65
Grade ≥G3 vs. <G3	**0.0069**	3.62	1.42–9.22	0.1247	2.38	0.79–7.18
IL-1β Low vs. high tumor expression (cut-off 160)	**0.0343**	2.27	1.06–4.84	**0.0460**	2.35	1.02–5.42
Age, year	**0.0047**	1.05	1.01–1.08	0.3421	1.02	0.98–1.06
Variable	Univariate analysis			Multivariate analysis		
OS	*p*	HR	95% CI	*p*	HR	95% CI
Pathological stage >T2 vs. T2	**0.0020**	2.79	1.45–5.33	0.1240	1.85	0.85–4.03
Nodal status N+ vs. N0	**0.0258**	1.85	1.08–3.17	0.1909	1.60	0.79–3.24
Metastasis status M+ vs. M0	**0.0036**	3.13	1.45–6.73	0.1148	2.12	0.83–5.37
Grade ≥G3 vs. < G3	**0.0173**	2.31	1.16–4.60	0.1611	1.94	0.77–4.88
IL-1β Low vs. high tumor expression (cut-off 160)	**0.0082**	2.54	1.27–5.06	0.0249	2.40	1.12–5.18
Age, year	**0.0031**	1.05	1.02–1.08	0.1274	1.03	0.99–1.07

3. Discussion

Intense research has been performed in recent years to identify potential biological markers predicting the clinical behavior of invasive BC [14,15]. This was followed by the significant improvement in our understanding of the molecular biology of this malignancy. Bladder tumors are highly heterogeneous, with frequent mutations in different signaling pathways, including cell cycle genes, receptors' tyrosine kinase, PI3K/AKT/mTOR, and chromatin regulatory gene mutations [16]. A number of studies have determined the important role of IL-1β in the occurrence and development of malignant tumors [12]. As a member of the IL-1 family of cytokines, IL-1β may act as a mediator of the inflammatory response in cancer, while involved in a variety of cellular activities, including cell proliferation, differentiation, and apoptosis. However, due to the complex repercussions on the course of cancer, both tumor-promoting and inhibitory functions of IL-1β have been described [17]; its specific role in BC carcinogenesis is yet to be determined. Additionally, IL1RA has been recognized as important factor in bladder carcinogenesis, and its high expression might be related to tumor aggressiveness [18].

In the present study, the expression patterns of IL-1β and IL1RA were assessed in invasive BC; a significant overexpression was observed when compared with normal bladder tissue. Moreover, the IL-1β expression was stronger in patients with adverse prognostic features (high-grade tumors (G3) and/or lymph-vascular invasion). Similar results have been reported in colorectal cancer [12], where the higher expression of IL-1β was associated with better OS and RFS, indicating a beneficial, prognostic role in rectal cancer. In the present analysis, we observed that only the IL-1β expression could be identified as an independent predictor of each RFS, CSS, and OS in a multivariate analysis. IL-1RA was also significantly overexpressed in invasive BC, as compared to benign urothelium, and a significant co-expression between IL-1β and IL-1RA was observed only in benign tissue. However, results showed that the lower expression of IL-1RA correlates with a high tumor grade (G3), which completely resembles results reported by Schneider et al. [18] indicating that adverse pathologic cancer characteristics relate to the low expression of IL-1RA. Moreover, authors emphasized the potential role of IL-1RA as a therapeutic target in patients with BC. This was examined in several trials, where the human competitive IL-1 inhibitor, Anakinra, decreased the proliferative rate of tumors refractory to standard therapies [19,20]. In this regard, our study revealed some promising results, since both adverse tumor features and the immune response of the host were obviously strongly influenced by the expression of the IL-1 axis.

Autophagy, the major endogenous pathway for degradation, delivery, and recycling of long-lived proteins and organelles, may be considered as a significant tumor suppressor [6]. According to recent data [21], the loss of the negative regulator of the PI3K autophagy pathway leads to more aggressive disease and metastasis, potentially through a stronger activation of the AKT mechanism. This was supported by our results, where the stronger expression of AKT is associated with a worse tumor grade, through a correlation with IL-1β. However, the interactions between the IL-1 axis and autophagy markers in invasive BC are still largely unknown. Noteworthy, recent evidence suggests that the IL-1β/1RA axis regulates cell proliferation, migration, clone formation, and apoptosis in vitro via autophagy mechanisms [22]. It seems that autophagy influences the IL-1β and pro-inflammatory cytokine secretion by macrophages. Interestingly, this effect could be both stimulatory and inhibitory [6,23]. All these data suggest that the inhibition of autophagy probably increases the IL-1β secretion through several mechanisms, with the sequestration into autophagosomes being the most dominant one [13,23]. This hypothesis is underlined by our results, where both IL-1β as well as the IL-1β/IL-1RA ratio correlated positively to the AKT expression in BC. It is, however, still unknown what the true prognostic potential of this relation is, despite clear evidence that AKT pathways influence the prognostic potential of the IL-1 axis in invasive BC. Proposed pathways and the interrelations of the analyzed biomarkers are illustrated in Figure 4.

We hypothesize that the IL-1β/1RA axis may express a positive correlation with Ki-67 in terms of tumor potential among patients with invasive BC. The observed interrelations of IL-1β expression with Ki-67 as a reliable marker for proliferation in BC suggests a putative approach for multivariate immunohistochemical profiles integrating, which improved the prognostic and presumably also the predictive accuracy by combining several markers [24]. Indeed, the IL-1β high expression showed a positive correlation with both the AKT and Ki-67 high staining in tumor cells, unlike IL-1 RA.

Although our analysis requires a large-scale validation, it suggests a potential mechanism-based selection strategy, in which patients with a higher expression of IL-1β in BC tissue may have a better immune response and probably a better response to cancer treatments. This attitude has been discussed in a report from Rebe et al. [17] with the conclusion that, although IL-1β has been recognized as a cancer promoter, it may also contribute the anti-tumor response, depending on the tumor stage and origin. These observations are in line with recent findings reporting a significant increase in the expression of the Kelch-like protein 14 antisense (Klhl14-AS) and Prader Willi/Angelman region RNA 5 (PAR5) levels in large- and high-grade cancer lesions [25]. These two Long noncoding RNAs seem to be severely downregulated in benign lesions, while in large malignant lesions, their levels are rescued, suggesting a potential tumor suppressor role in early phases of disease progression. We can speculate that the sample principle follows an IL-1β expression in BC, where the decrease in interleukin levels might promote early phases of tumor development [25]. However, this hypothesis must be confirmed in a large cohort with the support of a genetic analysis, in order to determine whether the high expression of IL-1β could be considered a marker of advanced disease and/or a better host anti-tumor response.

Limitations of the study lie in its retrospective character, single center design, and the reduced sample size. Inherent limitations of IHC may be observed in interobserver variability. Moreover, for the evaluation of the benign urothelium, corresponding tissue from tumor-free paraffin blocks from patients with BC was analyzed. Therefore, the expression pattern of biomarkers from totally healthy urothelium may differ from that observed in the present study. Finally, we must emphasize that a genetic profile of the IL-1 axis and correlation of IL-1β with an invasive capacity of BC cell lines was not included in our study, therefore reducing the validity of our results. However, the IHC analysis, as a pilot project for future work, clearly indicated that the IL-1 axis plays an important role in survival among patients with invasive BC.

4. Materials and Methods

4.1. Patient Cohorts

Based on an institutional database, patients undergoing radical cystectomy for invasive BC (urothelial carcinoma only) from February 1996 to December 2010 were identified. Patients with high-risk non-muscle invasive BC or a non-urothelial histology were not included in our analysis. Moreover, subjects younger than 18 years of age, with a history of previous radiotherapy, undergoing perioperative immunotherapy, suffering from recurrent or chronic infections, or any auto-immune disease have been excluded from the study (both groups). From a cystectomy specimen, a tissue microarray (TMA) was created that was based on the identification of representative sections of tumor and normal tissue in hematoxylin and eosin staining. Furthermore, corresponding histopathological benign urothelium tissue samples from surrounding areas (n = 22 and 39) were processed to a TMA and IHC stained for IL-1β, IL-1RA, AKT and Ki-67. The first group included 24 female and 78 male patients, whereas the second cohort consisted of 27 female and 70 males. The tumor grade and stage were assessed according to the WHO 2016 TNM classification system by at least two pathologists experienced in urogenital pathology [26]. Clinicopathological characteristics were recorded. The study was approved by the institutional review board of the University of Tubingen No. 279/2013BO2 from 1 March 2023.

4.2. Tissue Microarray (TMA) and Immunohistochemical (IHC) Staining

To obtain representative cores for the TMA construction, parallel sections stained by hematoxylin and eosin were used to identify a representative core position within the specimens. IHC staining was carried out according to the antibody manufacturers' instructions. Tissue slides were incubated overnight at 4 °C with a human IL-1 beta/IL-1F2 (AF-201, R&D systems Inc., Minneapolis, MN, USA) and IL-1 RA/IL-1F3 (AF-280, R&D systems, Minneapolis, MN, USA) polyclonal goat immunoglobulin, respectively [12,15], in dilutions 1:10 and 1:300, in real antibody diluent (DAKO, Glostrup, Denmark). After three more washing steps, a visualization was performed with Dako Liquid DAB-Substrat Chromogen System K3467 (DAKO, Glostrup, Denmark) and counterstaining with hematoxylin, as indicated by the manufacturer [27]. Two or more cores of every invasive BC and corresponding normal bladder tissue were integrated.

4.3. Immunohistochemical Protocol

TMAs were evaluated in a blinded manner by two independent reviewers, and divergent results were reevaluated. For the results, IL-1β and IL-RA cytoplasmatic cellular staining was scored using a 4-point scale (0, no staining; 1+, light staining at high magnification; 2+, intermediate staining; 3+, and dark staining of linear membrane at low magnification [22]. Expressions were then quantified by the histochemical scoring system/H-score 0–300 [28], and a quotient IL-1β/IL-1RA was built for further calculations. A microscopic analysis was performed at ×100 and ×400 magnifications. For AKT, epithelial zones were scored according to the intensity of staining of the cytoplasm, nucleus, or membrane, and the same scoring system was used [29]. Individual results were demonstrated as staining classes and compared to clinical and histopathological data of the second cohort, whereas the clinical course was evaluated in the first group. The Ki-67 score was expressed as the percentage of the number of immune-stained nuclei among the total number of nuclei of tumor cells, regardless of the immunostaining intensity. This counting was performed in three representative selected fields of the BC tissue section at ×400 magnification [30]. Ki-67 score ranged from 0–100%; its cut-off level was 15%, where the 'low immunoreactivity' was defined by nuclear staining of <15% and 'high' for staining ≥15% [24].

4.4. Patient Follow-up

In the conformation cohort, patient charts and physicians' records were reviewed to determine the clinical outcome. In general, patients were observed postoperatively at least every 3–4 months for the first year, semi-annually for the 2nd and 3rd year, and annually thereafter. The radiological imaging including the computed tomography was performed in all patients. The time until recurrence (RFS), cancer-specific survival (CSS), and overall survival (OS) were assessed [31].

4.5. Statistical Analysis

The expression of IL-1β, IL-1RA, AKT, and Ki-67 was correlated with clinicopathologic parameters at the time of cystectomy by Wilcoxon Kruskal–Wallis tests, Chi-square tests, or linear regression analyses, dependent on the variable's category. Kaplan–Meier analyses were used to estimate recurrence-free (RFS), cancer-specific (CSS) and overall survival (OS) log-rank test. Uni- and multivariate Cox proportional hazard analyses were performed to assess the impact of IL-1β and IL-1RA together with other relevant clinical and pathologic variables. *p*-values <0.05 were considered significant. JMP, version 16.2 (SAS Institute, Cary, NC, USA) was used for statistics.

5. Conclusions

We found that the high expression level of IL-1β correlates with RFS, CSS, and OS in bladder cancer and is an independent predictor of the outcome in the multivariate analysis. Moreover, both IL-1β and the IL-1β/IL-1RA ratio show a positive correlation to the AKT and Ki-67 expression, indicating an involvement in autophagy and high proliferative activity. Combined analysis of these markers may contribute to clinical decision making when defining therapeutic strategies for invasive BC.

Author Contributions: M.V., Data curation, Analysis, Writing, and Original draft preparation. J.M.C. and S.J., Data curation and Analysis. J.H., T.T. and T.L., Data acquisition, Supervision, Reviewing, and Editing. I.T. and A.S., Supervision. S.R., Project development, Analysis, Writing, Reviewing, and Editing. All authors have read and agreed to the published version of the manuscript.

Funding: Vukovic Marko is supported by EUSP scholarship grant of the European Association of Urology (S-2021-0013).

Institutional Review Board Statement: The study was approved by the institutional review board of the University of Tubingen No. 279/2013BO2 on 1 March 2023.

Informed Consent Statement: Informed consent was obtained from all subjects involved in the study.

Data Availability Statement: The data from this study are available by request from the corresponding author. The data are not publicly available due to ethical restrictions (IRB statement).

Acknowledgments: We thank Sieglinde Baisch and Irini Grigoriadou from Eberhard-Karls University of Tuebingen for the scientific support.

Conflicts of Interest: The authors declare no conflicts of interest.

References

1. Wang, J.Z.; Zhu, W.; Han, J.; Yang, X.; Zhou, R.; Lu, H.C.; Yu, H.; Yuan, W.B.; Li, P.C.; Tao, J.; et al. The role of the HIF-1α/ALYREF/PKM2 axis in glycolysis and tumorigenesis of bladder cancer. *Cancer Commun.* **2021**, *41*, 560–575. [CrossRef]
2. Huang, W.; Zhang, C.; Li, Y.; Zha, H.; Zhou, X.; Fu, B.; Guo, J.; Wang, G. IGF2BP3 facilitates cell proliferation and tumorigenesis via modulation of JAK/STAT signalling pathway in human bladder cancer. *J. Cell. Mol. Med.* **2020**, *24*, 13949–13960. [CrossRef]

3. Rausch, S.; Hennenlotter, J.; Teepe, K.; Kuehs, U.; Aufderklamm, S.; Bier, S.; Mischinger, J.; Gakis, G.; Stenzl, A.; Schwentner, C.; et al. Muscle-invasive bladder cancer is characterized by overexpression of thymidine kinase 1. *Urol. Oncol. Semin. Orig. Investig.* **2015**, *33*, 426.e21–426.e29. [CrossRef]
4. Xia, H.; Green, D.R.; Zou, W. Autophagy in tumour immunity and therapy. *Nat. Rev. Cancer* **2021**, *21*, 281–297. [CrossRef]
5. Vetterlein, M.W.; Roschinski, J.; Gild, P.; Marks, P.; Soave, A.; Doh, O.; Isbarn, H.; Höppner, W.; Wagner, W.; Shariat, S.F.; et al. Impact of the Ki-67 labeling index and p53 expression status on disease-free survival in pT1 urothelial carcinoma of the bladder. *Transl. Androl. Urol.* **2017**, *6*, 1018–1026. [CrossRef]
6. Liu, G.H.; Zhong, Q.; Ye, Y.L.; Wang, H.B.; Hu, L.J.; Qin, Z.K.; Zeng, M.S.; Zeng, B.H. Expression of beclin 1 in bladder cancer and its clinical significance. *Int. J. Biol. Markers* **2013**, *28*, 56–62. [CrossRef]
7. Kim, H.; Lee, S.J.; Lee, I.K.; Min, S.C.; Sung, H.H.; Jeong, B.C.; Lee, J.; Park, S.H. Synergistic Effects of Combination Therapy with AKT and mTOR Inhibitors on Bladder Cancer Cells. *Int. J. Mol. Sci.* **2020**, *21*, 2825. [CrossRef]
8. Matsumoto, H.; Munemori, M.; Shimizu, K.; Fujii, N.; Kobayashi, K.; Inoue, R.; Yamamoto, Y.; Nagao, K.; Matsuyama, H. Risk stratification using Bmi-1 and Snail expression is a useful prognostic tool for patients with upper tract urothelial carcinoma. *nt. J. Urol.* **2016**, *23*, 1030–1037.
9. Wang, F.; Liu, J.; Chen, X.; Zheng, X.; Qu, N.; Zhang, B.; Xia, C. IL-1β receptor antagonist (IL-1Ra) combined with autophagy inducer (TAT-Beclin1) is an effective alternative for attenuating extracellular matrix degradation in rat and human osteoarthritis chondrocytes. *Arthritis Res. Ther.* **2019**, *21*, 171. [CrossRef]
10. De Luca, A.; Smeekens, S.; Casagrande, A.; Iannitti, R.; Conway, K.L.; Gresnigt, M.S.; Begun, J.; Plantinga, T.S.; Joosten, L.A.; van der Meer, J.W.; et al. IL-1 receptor blockade restores autophagy and reduces inflammation in chronic granulomatous disease in mice and in humans. *Proc. Natl. Acad. Sci. USA* **2014**, *111*, 3526–3531. [CrossRef]
11. Crişan, T.O.; Plantinga, T.S.; van de Veerdonk, F.L.; Farcaş, M.F.; Stoffels, M.; Kullberg, B.-J.; van der Meer, J.W.M.; Joosten, L.A.; Netea, M.G. Inflammasome-independent modulation of cytokine response by autophagy in human cells. *PLoS ONE* **2011**, *6*, e18666. [CrossRef]
12. Semilia, M.; Hennenlotter, J.; Pavone, C.; Bischoff, T.; Kühs, U.; Gakis, G.; Bedke, J.; Stenzl, A.; Schwentner, C.; Todenhöfer, T. Expression patterns and prognostic role of transketolase-like 1 in muscle-invasive bladder cancer. *World J. Urol.* **2015**, *33*, 1403–1409. [CrossRef]
13. Wilson, F.; Joseph, N.; Choudhury, A. Biomarkers in muscle invasive bladder cancer. *Adv. Clin. Chem.* **2022**, *107*, 265–297. [PubMed]
14. Audenet, F.; Attalla, K.; Sfakianos, J.P. The evolution of bladder cancer genomics: What have we learned and how can we use it? *Urol. Oncol.* **2018**, *36*, 313–320. [CrossRef]
15. Chen, Y.; Yang, Z.; Deng, B.; Wu, D.; Quan, Y.; Min, Z. Interleukin 1β/1RA axis in colorectal cancer regulates tumor invasion, proliferation and apoptosis via autophagy. *Oncol. Rep.* **2020**, *43*, 908–918. [CrossRef]
16. Rébé, C.; Ghiringhelli, F. Interleukin-1β and Cancer. *Cancers* **2020**, *12*, 1791. [CrossRef]
17. Schneider, L.; Liu, J.; Zhang, C.; Azoitei, A.; Meessen, S.; Zheng, X.; Cremer, C.; Gorzelanny, C.; Kempe-Gonzales, S.; Brunner, C.; et al. The Role of Interleukin-1-Receptor-Antagonist in Bladder Cancer Cell Migration and Invasion. *Int. J. Mol. Sci.* **2021**, *22*, 5875. [CrossRef]
18. Waugh, J.; Perry, C.M. Anakinra: A review of its use in the management of rheumatoid arthritis. *BioDrugs* **2005**, *19*, 189–202. [CrossRef]
19. Lust, J.A.; Lacy, M.Q.; Zeldenrust, S.R.; Dispenzieri, A.; Gertz, M.A.; Witzig, T.E.; Kumar, S.; Hayman, S.R.; Rusell, S.R.; Buadi, F.K.; et al. Induction of a chronic disease state in patients with smoldering or indolent multiple myeloma by targeting interleukin 1{beta}-induced interleukin 6 production and the myeloma proliferative component. *Mayo Clin. Proc.* **2009**, *84*, 114–122. [CrossRef]
20. Minoli, M.; Kiener, M.; Thalmann, G.N.; Kruithof-de Julio, M.; Seiler, R. Evolution of Urothelial Bladder Cancer in the Context of Molecular Classifications. *Int. J. Mol. Sci.* **2020**, *21*, 5670. [CrossRef]
21. Yao, S.J.; Ma, H.S.; Liu, G.M.; Gao, Y.; Wang, W. Increased IL-1α expression is correlated with bladder cancer malignant progression. *Arch. Med. Sci.* **2020**, *19*, 160–170. [PubMed]
22. Iula, L.; Keitelman, I.A.; Sabbione, F.; Fuentes, F.; Guzman, M.; Galletti, J.G.; Gerber, P.P.; Ostrowski, M.; Geffner, J.R.; Jancic, C.C.; et al. Autophagy Mediates Interleukin-1β Secretion in Human Neutrophils. *Front. Immunol.* **2018**, *9*, 269. [CrossRef] [PubMed]
23. Wu, N.; Zhu, Y.; Xu, X.; Zhu, Y.; Song, Y.; Pang, L.; Chen, Z. The anti-tumor effects of dual PI3K/mTOR inhibitor BEZ235 and histone deacetylase inhibitor Trichostatin A on inducing autophagy in esophageal squamous cell carcinoma. *J. Cancer.* **2018**, *9*, 987–997. [CrossRef] [PubMed]
24. Bertz, S.; Otto, W.; Denzinger, S.; Wieland, W.F.; Burger, M.; Stöhr, R.; Link, S.; Hofstädter, F.; Hartmann, A. Combination of CK20 and Ki-67 immunostaining analysis predicts recurrence, progression, and cancer-specific survival in pT1 urothelial bladder cancer. *Eur. Urol.* **2014**, *65*, 218–226. [CrossRef]
25. Spirito, L.; Maturi, R.; Credendino, S.C.; Manfredi, C.; Arcaniolo, D.; De Martino, M.; Esposito, F.; Napolitano, L.; Di Bello, F.; Fusco, A.; et al. Differential Expression of LncRNA in Bladder Cancer Development. *Diagnostics* **2023**, *13*, 1745. [CrossRef] [PubMed]
26. Humphrey, P.A.; Moch, H.; Cubilla, A.L.; Ulbright, T.M.; Reuter, V.E. The 2016 WHO Classification of Tumours of the Urinary System and Male Genital Organs-Part B: Prostate and Bladder Tumours. *Eur Urol.* **2016**, *70*, 106–119. [CrossRef] [PubMed]

27. John, A.; Günes, C.; Bolenz, C.; Vidal-Y-Sy, S.; Bauer, A.T.; Schneider, S.W.; Gorzelanny, C. Bladder cancer-derived interleukin-1 converts the vascular endothelium into a pro-inflammatory and pro-coagulatory surface. *BMC Cancer* **2020**, *2*, 1178. [CrossRef]
28. Andersen, N.L.; Brügmann, A.; Lelkaitis, G.; Nielsen, S.; Lippert, M.F.; Vyberg, M. Virtual Double Staining: A Digital Approach to Immunohistochemical Quantification of Estrogen Receptor Protein in Breast Carcinoma Specimens. *Appl. Immunohistochem. Mol. Morphol.* **2018**, *26*, 620–626. [CrossRef]
29. Le Page, C.; Koumakpayi, I.H.; Alam-Fahmy, M.; Mes-Masson, A.-M.; Saad, F. Expression and localisation of Akt-1, Akt-2 and Akt-3 correlate with clinical outcome of prostate cancer patients. *Br. J. Cancer* **2006**, *94*, 1906–1912. [CrossRef]
30. Aman, N.A.; Doukoure, B.; Koffi, K.D.; Koui, B.S.; Traore, Z.C.; Kouyate, M.; Toure, I.; Effi, A.B. Immunohistochemical Evaluation of Ki-67 and Comparison with Clinicopathologic Factors in Breast Carcinomas. *Asian Pac. J. Cancer Prev.* **2019**, *20*, 73–79. [CrossRef]
31. Liu, Y.; Tang, Y.; Wie, Y.W. What is the difference between overall survival, recurrence-free survival and time-to-recurrence? *Br. J. Surg.* **2020**, *107*, e634. [PubMed]

International Journal of *Molecular Sciences*

MDPI

Article

MMP-14 Exhibits Greater Expression, Content and Activity Compared to MMP-15 in Human Renal Carcinoma

Grzegorz Młynarczyk [1,*], Anna Tokarzewicz [2], Monika Gudowska-Sawczuk [3], Barbara Mroczko [3,4], Vojtěch Novák [5], Adam Novák [5], Przemysław Mitura [6] and Lech Romanowicz [2]

1 Department of Urology, Medical University of Białystok, 15-276 Białystok, Poland
2 Department of Medical Biochemistry, Medical University of Białystok, ul. Mickiewicza 2, 15-089 Białystok, Poland; anna.tokarzewicz@umb.edu.pl (A.T.); lech.romanowicz@umb.edu.pl (L.R.)
3 Department of Biochemical Diagnostics, Medical University of Białystok, 15-269 Białystok, Poland; monika.gudowska-sawczuk@umb.edu.pl (M.G.-S.); barbara.mroczko@umb.edu.pl (B.M.)
4 Department of Neurodegeneration Diagnostics, Medical University of Bialystok, 15-269 Białystok, Poland
5 Department of Urology, Charles University and University Hospital Motol, 150 06 Praha, Czech Republic; vojtech.novak.lf2@gmail.com (V.N.); adamnovak1229@gmail.com (A.N.)
6 Department of Urology and Oncological Urology, Medical University of Lublin, 20-059 Lublin, Poland; przemyslawmitura@umlub.pl
* Correspondence: grzegorz.mlynarczyk@umb.edu.pl; Tel.: +48-85-748-55-79; Fax: +48-85-748-55-78

Abstract: Membrane-type metalloproteinases (including MMP-14 and MMP-15) are enzymes involved in the degradation of extracellular matrix components. In cancer, they are involved in processes such as cellular invasion, angiogenesis and metastasis. Therefore, the aim of this study was to evaluate the expression, content and activity of MMP-14 and MMP-15 in human renal cell carcinoma. Samples of healthy kidney tissue (n = 20) and tissue from clear-cell kidney cancer (n = 20) were examined. The presence and contents of the MMPs were assessed using Western blot and ELISA techniques, respectively. Their activity—both actual and specific—was evaluated using fluorimetric analysis. Both control and cancer human kidney tissues contain MMP-14 and MMP-15 enzymes in the form of high-molecular-weight complexes. Moreover, these enzymes occur in both active and latent forms. Their content in cancer tissues is very similar, but with a noteworthy decrease in content with an increase in the kidney cancer grade for both membrane-type metalloproteinases. Even more notable is the highest content of the investigated enzymes represented by MMP-14 in the control tissues. Considering the actual and specific activity outcomes, MMP-14 dominates over MMP-15 in all of the investigated tissues. Nevertheless, we also noted a significant enhancement of the activity of both metalloproteinases with an increase in the grade of renal cancer. The expression and activity of both enzymes were detected in all examined renal cancer tissues. However, our findings suggest that transmembrane metalloproteinase 14 (MMP-14) plays a much more significant and essential role than MMP-15 in the studied renal carcinoma tissues. Therefore, it seems that MMP-14 could be a promising target in the diagnosis, prognosis and therapy of renal cell carcinoma.

Keywords: cancer; renal carcinoma; transmembrane metalloproteinases; MMP-14; MMP-15

1. Introduction

The kidneys are even, parenchymal organs located in the retroperitoneal space of the human abdominal cavity. Their main functions are the excretion of metabolic products and the regulation of the components, volume and pH of body fluids. They also have metabolic and endocrine functions [1–3]. The structure of the extracellular matrix has a significant impact on their proper functioning. Its components, including collagen (the main extracellular protein), glycoproteins and elastin molecules, form a complex network that also interacts with surrounding cells.

The ECM is present in the renal cortex in anatomically distinct areas and has varied functions, depending on its molecular components: glomeruli, tubulointerstitium and in vessels. Besides ECM remodeling, MMPs take part in several other processes, such as basement membrane destruction, angiogenesis, cell migration and cell apoptosis [4–6]. The group of membrane-type metalloproteinases consists of six enzymes: MMP-14, MMP-15, MMP-16, MMP-24, MMP-17 and MMP-25. Various components of the extracellular matrix, such as collagen types I, II and III; aggrecan; elastin; fibronectin; gelatin; laminin and MMP-2 and -13, are degraded by these enzymes. At the same time, they are simultaneously involved in signal transfer at the cellular level [7–9].

MMP-14, the first discovered membrane-type metalloproteinase, became the most frequently selected subject of MMP research. The main substrates degraded by MMP-14 are type I collagen and pro-MMP-2. Collagen II and III can also be degraded by this enzyme, but to a lesser extent [10]. The enzyme also degrades gelatin and the hyaluronate receptor [11]. To activate MMP-2, MMP-14 forms a ternary complex on the cell surface with TIMP-2 (Tissue Inhibitor of Metalloproteinases 2) and with MMP-2 [12–14]. It plays a very important role in angiogenesis. Furthermore, MMP-14 promotes cellular invasion in cancerogenic processes, metastasis, angiogenesis, wound healing, atherosclerosis and rheumatoid arthritis [15,16]. The overexpression of MMP-14 has been observed in kidney fibrosis [17]. Moreover, MMP-14 expression varies with the tumor type and is high in melanomas, mesenchymal tumors and brain tumors [18], as well as in liver tumors and breast cancer [19,20]. Some authors have suggested that MMP-14 can be used to predict cancer prognosis [21].

It has been shown that the active form of MMP-15 plays a role in breast, prostate and colon cancerogenic processes. It degrades only type 1 collagen; however, its specific activity outcomes showed only 1/100 that of MMP-14 activity. Moreover, it was demonstrated that MMP-15 may promote cell invasion into basement membrane matrices [22–24].

Transmembrane metalloproteinases MMP-14, MMP-15 and MMP-16 have been found to be significantly expressed in tumor-transformed tissue in renal cell carcinoma and to be present in urine samples from the same patients [25]; however, no reports have been published on their expression in renal cell carcinoma.

Transmembrane metalloproteinases play an important part in renal cancer; however, their significance is not entirely clear. Therefore, the expression, content and activity of MMP-14 and MMP-15 in cancerous tissue in relation to healthy human kidney tissue were investigated.

2. Results

2.1. MMP-14 and MMP-15 Content in Tissue Samples

Based on the measurements described in Section 4.3, considerable differences in the content (mg/kg of total protein content) of both investigated MMPs in control and renal cell carcinoma samples were noted.

MMP-14 was present in an extremely high quantity in control tissue samples, at approximately 1.1 g of MMP-14/kg of total protein content, regardless of the source of control material (control G2 or control G3). The G2 and G3 tumor samples contained substantially lower amounts of MMP-14. The G2 tumor samples contained 14.87, whilst the control contained 1057 mg of MMP-14/kg of total protein content. Even lower quantities were found in the G3 tumor, as compared to the control tissue (12.93 vs. 1119.8 mg of MMP-14/kg of total protein content). Thus, the progress in tumor invasiveness from tumor grade G2 to G3 involved a rather small decrease in the MMP-14.

The approximate content of MMP-15 in control samples was 19.48 (control of G2) and 15.69 (control of G3) mg of MMP-15/kg of total protein content. The G2 and G3 tumor samples contained smaller amounts of MMP-15. There was almost 2 mg less of the enzyme in the G2 tumor. An even lower value (approximately 4 mg less of the enzyme) was found in the G3 tumor, compared to the respective control tissue samples. Such progress in tumor invasiveness from tumor grade G2 to G3 involved a significant decrease in the MMP-15 content. The MMP content results are provided in Figure 1.

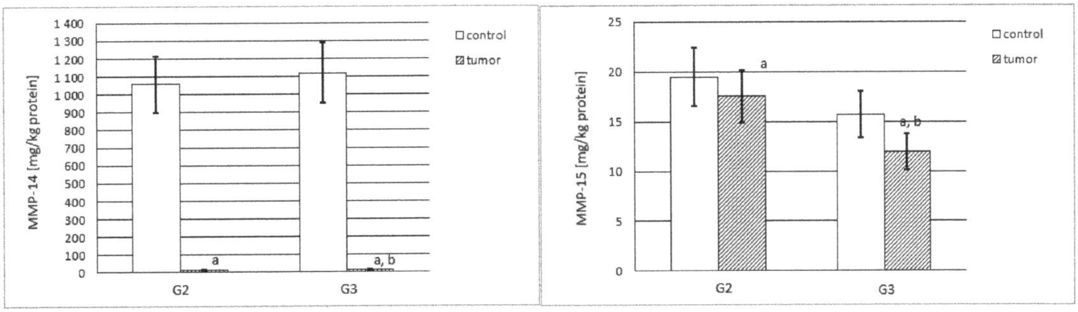

Figure 1. Content of MMP-14 and MMP-15 in control samples and renal cell carcinoma (G2 and G3) samples. MMP-14: a—$p < 0.001$ control vs. renal cell carcinoma; b—$p < 0.05$ G2 vs. G3; MMP-15: a—$p < 0.05$ control vs. renal cell carcinoma; b—$p < 0.001$ G2 vs. G3.

2.2. Western Blot Analysis of MMP-14 and MMP-15

2.2.1. Expression of MMP-14

To determine the expression of MMP-14 in renal cell carcinoma samples and the control samples, Western Blot analysis was performed under both non-reducing and reducing condition. Non-reducing conditions allow the separation of native, free forms of proteins as well as their complexes. The use of a reducing agent breaks the disulphide bonds present in proteins and complexes. It shows the presence of subunits or protein fragments, if they are linked by disulphide bridges. It is also possible that the separation of protein subunits results in the disappearance of protein binding sites that allowed the formation of high-molecular-weight complexes. This causes changes in the Western blot obtained results.

Figure 2 presents the results of a representative analysis, from which it can be seen that all of the tested samples contained MMP-14 in the form of high-molecular-weight complexes, as indicated by an intense band of approximately 202 kDa under non-reducing conditions (lanes 1–4). In addition, both control samples (lane 1: G2 control, and lane 3: G3 control) showed an intense band of approximately 48 kDa in molecular weight, while both tumor samples (lane 2: G2, and lane 4: G3) showed a non-intensive band of the discussed mass. A band of approximately 40 kDa was present in both control samples and renal cell carcinoma samples. No other bands of lower masses were found under non-reducing conditions.

Reduction of the disulphide bridge with 2-mercaptoethanol resulted in the disappearance of high-molecular-weight complexes in all of the tested samples (lanes 5–8). Both control and renal cell carcinoma samples showed bands with approximately 60, 55, 48, and 40 kDa molecular weight. All four samples also demonstrated a non-intensive band with a molecular weight of 30 kDa (lanes 5–8; Figure 2).

2.2.2. Expression of MMP-15

Figure 3 shows that applying even a low mass of protein (20 micrograms) to the gel resulted in the appearance of intense bands with molecular masses of approximately 202 kDa and 48 kDa in all of the tested samples, without any other bands of lower masses, under non-reducing conditions (lanes 1–4). After the reduction of disulphide bonds, only two bands (approximately 50 kDa and 30 kDa) can be seen on the Western Blot gel, which was separated and present in all the examined tissues (lanes 5–8).

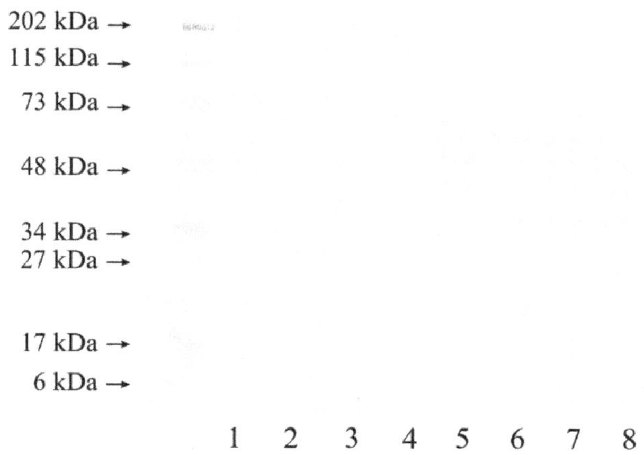

Figure 2. Representative results of Western immunoblot analysis of MMP-14 in renal cell carcinoma. Non-reducing conditions: lane: 1—control sample for grade G2 tumor; 2—grade G2 tumor; 3—control sample for grade G3 tumor; 4—grade G3 tumor. Reducing conditions: lane: 5—control sample for grade G2 tumor; 6—grade G2 tumor; 7—control sample for grade G3 tumor; 8—grade G3 tumor; 20 µg of protein was applied to the gel.

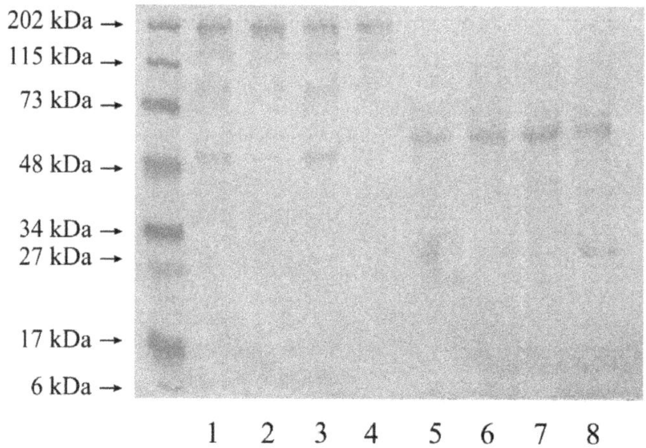

Figure 3. Representative results of Western immunoblot analysis of MMP-15 in renal cell carcinoma. Non-reducing conditions: lane: 1—control sample for grade G2 tumor; 2—grade G2 tumor; 3—control sample for grade G3 tumor; 4—grade G3 tumor. Reducing conditions: lane: 5—control sample for grade G2 tumor; 6—grade G2 tumor; 7—control sample for grade G3 tumor; 8—grade G3 tumor; 20 µg of protein was applied to the gel.

2.3. Activities of MMP-14 and MMP-15

The actual and specific activities of both MMPs were evaluated through fluorimetric assay with an oligopeptide as a substrate. Each of the discussed enzymes was isolated on a microplate, which was pre-coated with a suitable and selective antibody. After immobilization of the enzymes, all remaining proteins were washed out from the microplate. The quantity of enzyme degradation product was determined after the addition of fluorogenic substrate.

The actual activity of membrane-type matrix metalloproteinases allows for the comparison of the activity of all the enzymes tested with each other (with the total protein content of the tissue extract as a common denominator). Moreover, the specific activity allowed for determination of what part of the enzyme was in the active form, without an inhibitor bound to the enzyme's active site.

2.3.1. Actual and Specific Activity of MMP-14

As shown in Figure 4A, similar actual activity of MMP-14 (approximately 15 nkatal/kg of protein content) was found in all control samples. The determined actual activity was more than two times lower in G2 tumor samples and almost four times lower in G3 tumor samples when compared to corresponding control samples; therefore, the actual activity significantly decreased with increasing grade of renal cell carcinoma.

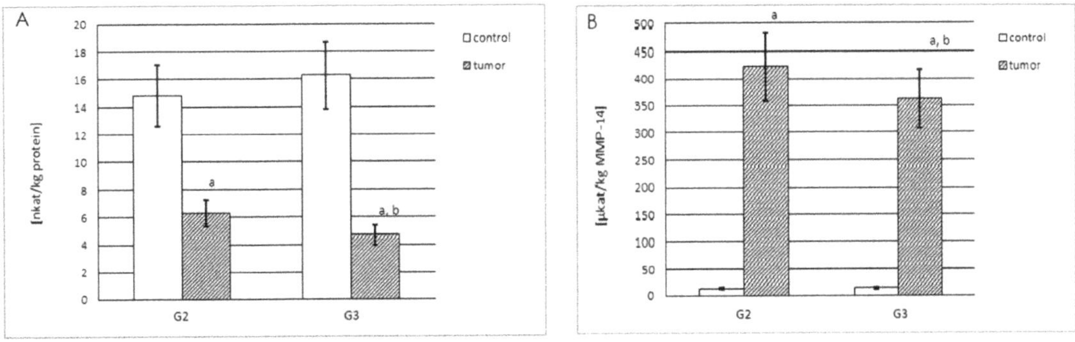

Figure 4. (**A**) Actual activity of MMP-14 (nkatal/kg of total protein content) in control samples (total n = 20), and grade G2 (n = 10) and G3 (n = 10) renal cell carcinoma samples; a, $p < 0.001$ cancer vs. respective control sample; b, $p < 0.001$ grade G3 vs. G2 renal cell carcinoma samples. (**B**) Specific activity of MMP-14 (μkatals/kg of the MMP-14) in the same samples (each group n = 10); a, $p < 0.001$ cancer vs. respective control sample; b, $p < 0.05$ grade G3 vs. G2 renal cell carcinoma samples.

In all of the tested samples, the content of MMP-14 was measured beforehand, which enabled the possibility of determining the specific activity of MMP-14 (μkatal/kg of MMP-14). From Figure 4B, it can be seen that both control samples were characterized by very low and similar specific activity, approximately 14 μkatals/kg of MMP-14. In addition, specific activity in the G2 and G3 tumor samples was nearly 30 times and more than 25 times higher in comparison to the control samples, respectively.

2.3.2. Actual and Specific Activity of MMP-15

The highest actual MMP-15 activity (pkatals/kg of total protein content) was found in the grade G2 renal cell carcinoma samples (Figure 5A). The increase in the grade of renal cell carcinoma was accompanied by a decrease in enzyme activity; however, it was still significantly higher than in the corresponding control samples. The actual activity was 35% higher in G3 and 45% higher in G2 tumor samples, compared to the corresponding control samples.

The specific activity (μkatals/kg of MMP-15) of this enzyme was also determined. Figure 5B shows that this type of activity was more than 1.5 times higher in G2 tumor samples, compared with the respective control samples. The G3 tumor samples demonstrated the highest specific activity of MMP-15—approximately 1.8 times higher than in the respective control sample and over 22% higher than in the G2 tumor samples.

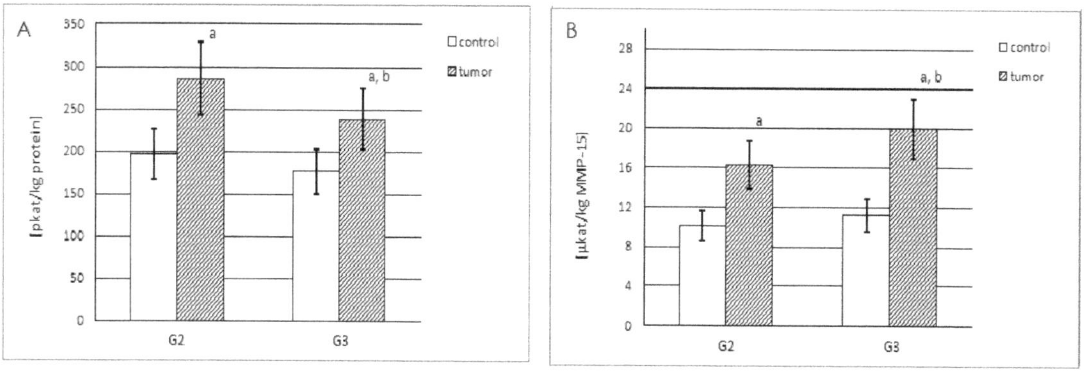

Figure 5. (**A**) Actual activity of the MMP-15 (pkatals/kg of total protein content) in control samples (total n = 20), and grade G2 (n = 10) and G3 (n = 10) renal cell carcinoma samples; a, $p < 0.001$ cancer vs. respective control samples; b, $p < 0.05$ grade G3 vs. G2 renal cell carcinoma samples. (**B**) Specific activity of the MMP-15 (µkatals/kg of the MMP-15) in the same samples (each group n = 10); a, $p < 0.001$ cancer vs. respective control samples; b, $p < 0.05$ grade G3 vs. G2 renal cell carcinoma samples.

3. Discussion

A small amount of extracellular matrix has been observed in the human kidney. In addition to collagen, the ECM contains many other important components, such as elastin, proteoglycans and structural glycoproteins. The basement membrane, which separates epithelial tissue from connective tissue, is a special layer of the ECM. This membrane is mainly composed of type IV collagen and glycoproteins [26–28]. Many researchers have investigated the characteristics of human renal cell carcinoma [29–32]. The focus of the study described in this article was to evaluate the expression, content and activity of the MMPs involved in the degradation of ECM components in human kidney tissues.

Membrane-type metalloproteinases are represented by six enzymes. We compared MMP-14 (membrane-type MMP 1) and MMP-15 (membrane-type MMP 2) in human kidney tumor to those in healthy parts of the same organ, which were used as a control material as ethical concerns excluded the possibility of taking a kidney sample from healthy donors. Post-mortem human kidney collection entails a substantial change in protein content and is possible after several hours. After such time, the tissue degradation process is usually advanced. According to Lee et al. [33], all of the enzymes degrading the extracellular matrix are only active components. The content of protein is much lower, when compared with the levels found in tissue that has been newly taken from a living organism.

ELISA was used to measure the total content of the tested MMP-14 and MMP-15 in kidney samples. We observed that human control kidney tissue showed higher amounts of both MMPs, compared to tumor tissue. Based on this, it can be speculated that both enzymes may play an important role in the restoration of the extracellular matrix in healthy kidneys. It is worth noting that the highest content was demonstrated in both control tissues for MMP-14, while a very similar amount of both enzymes was found in both grades of cancer. The above outcomes demonstrate that the synthesis and secretion of investigated metalloproteinases from cells were not inhibited. Moreover, there are differences between malignant and normal renal cells in terms of enzyme secretion outside the cell. The synthesis of both transmembrane metalloproteinases may be at a similar level for both kinds of cells, but their secretion to the extracellular space is limited by cancer cells. It is common knowledge that protein turnover is much faster inside the cell than outside it [34]. MMPs, similar to all other matrix metalloproteinases, are extracellular enzymes and, therefore, their lower content in the tumor tissue may be explained by the decrease in enzyme secretion from malignant cells.

On the basis of the actual activity of the tested MMPs (katal per kg of total protein content), the activity of the two analyzed MMPs was compared in different tissue samples. Based on the results, it can be concluded that MMP-14 is significantly less active in tumor tissue than in control tissue; at the same time, it is also approximately 70 times more active in healthy human kidney samples than MMP-15. Based on this, it seems that MMP-14 is significantly involved in maintaining ECM homeostasis in healthy kidneys.

MMP-15 demonstrated a significantly higher activity in both cancer grades, as compared to the control material. However, MMP-14 demonstrated a much higher activity in both grades of kidney cancer, when compared to MMP-15. Such a difference in the actual activity of both investigated enzymes may provide further proof explaining the significant role of MMP-14 in carcinogenic processes, as well as indicating that it does not reduce the role of MMP-15.

Furthermore, the activities of MMP-14 and MMP-15 in the cells of both tumor grades were determined. However, there was no relation between enzyme expression/activity and cancer progression.

Based on the specific activity results for each of the MMPs tested (μkatal per kg of enzyme), it can be observed that part of the enzyme tested was present in the active form, without an inhibitor bound to the active site of the enzyme. Both examined enzymes showed a similar specific activity in the control tissue. Furthermore, both enzymes demonstrated a significantly higher activity in the cancerous material compared to the control tissue. In addition, a significantly higher specific activity of MMP-14 in tumor tissue was determined, compared to MMP-15. This appears to be related to the wide range of proteins degraded by MMP-14, including pro-MMP-2 activation [12–14]. In summary, it can be concluded that most MMP-14, unlike MMP-15, remains in an active form without the inhibitory effect of TIMPs. It seems that the MMP-14 enzyme showed greater enzymatic activity compared to MMP-15. Other authors have also demonstrated the over-expression of MMP-14 at its mRNA stage in various types of tumors [18–20].

As shown by the Western blot analysis results, both studied enzymes were present in kidney tissues primarily in the form of high-molecular-weight complexes. This is due to the fact that the extracellular matrix is a tissue structure in which different proteins interact with each other. In many cases, these interactions do not affect enzyme activity. The described complexes disappear as a result of the reduction of disulphide bridges. This means that the bonds connecting all components are weak and non-covalent.

In all tested tissue samples, MMP-14 was present in a free active form. This is evidenced by the presence of a band with a molecular weight of approximately 50 kDa, both before and after disulphide bond reduction. The obtained results also demonstrated the presence of MMP-15 in free active form in all samples, due to a clearly visible band with a molecular weight of 50 kDa, both before and after the reduction of disulphide bonds.

The results of other studies have demonstrated that the transmembrane metalloproteinases examined in this study were significantly expressed during the course of renal cell carcinoma in tumor-lesioned tissue and were present in urine samples from the same patients [25]. However, published reports on the expression of these enzymes in renal cell carcinoma are still lacking.

Even though the content of both investigated membrane-type metalloproteinases in renal cancer was similar, the activity of MMP-14 was much higher in both tumor phases compared to that of MMP-15. This may mean that the catalytic ability of MMP-14 is much higher in comparison to MMP-15. Despite their similar contents, MMP-14 seems to be the dominant MMP in kidney cancer. On the other hand, the differences between outcomes for both enzymes may point to their specific involvement in a particular growth period and in cancer differentiation. Furthermore, the above outcomes show that cancerous cells of grade G2 and G3 kidney tumors may enhance the activity of transmembrane metalloproteinases. Based on the abovementioned similar content of both enzymes and the lower specific activity of MMP-15 in comparison with MMP-14, it can be stated that many more molecules of MMP-15 in an inactive form were present. Such differences in the activity of both MMPs proved that they participated in the opposite way in different phases of ECM remodeling, and at different stages of tumor development. This indicates differences in regulating the expression and activation of the examined matrix metalloproteinases in human kidney tumors. Due to the fact that the control sample was taken from the same kidney, it is possible that there was a carcinogenic influence on the metabolism of the entire organ. It may be assumed that the initiation of ECM component degradation in the extracellular matrix is the most important stage in the process of tumor growth. The tissue material was used, instead of urine or blood serum of the same patient, in order to better reflect the roles of membrane-type metalloproteinases in tissue metabolism.

4. Materials and Methods

The study protocol was approved by the Bioethical Committee of the Medical University of Bialystok.

4.1. Tissue Material

The sample was composed of patients with clear-cell kidney cancer of grade G2 or G3 according to the ISUP/WHO malignancy scale. All patients underwent radical surgical treatment, namely open nephrectomy. A total of 20 patients (6 females and 14 males) aged between 48 and 78 years (mean age, 61 years) were selected for this study. The sample material included specimens of primary tumors collected during the surgery. Patients with cancers of grades G2 (n = 10) and G3 (n = 10) were selected as participants for this study. Tissue specimens of the same kidney, but always located opposite to the tumor, were collected as a comparative material. Any carcinogenetic lesion was macroscopically visible in the control tissue. All samples taken were washed using 0.9% NaCl solution, weighed, portioned and stored at −70 °C during the entire period of analysis. The surgical procedures were performed at the Department of Urology of the University Hospital in Bialystok.

4.2. Specimen Preparation

A 0.9% NaCl solution was used to wash the tissue samples. The tissues were then cut into small pieces, which were used to prepare the homogenate. Tris/HCl 0.05 M buffer (pH = 7.4) was used to prepare tissue homogenates (20%, *w*/*v*). A knife homogenizer (20,000 rpm, 3 × 30 s, 0 °C) was used for homogenization. In addition, homogenates were subjected to ultrasonification (20 kHz, 3 × 15 s, 0 °C). Then, centrifugation was performed at 10,000× *g* for 30 min at 4 °C. The supernatants (final tissue samples) obtained after centrifugation were collected, divided, and stored at −70 °C for further use.

4.3. MMP Contents

An MMP-14 ELISA kit provided by Sigma-Aldrich (Saint Louis, MO, USA; cat. no. RAB0362) and MMP-15 ELISA kit provided by Cloud-Clone Corp. (Katy, TX, USA; cat. no. SEA099Hu) were used to determine the contents of MMP-14 and MMP-15, respectively, in the test samples. Both quantitative assays were performed according to the manufacturer's instructions.

4.4. MMP Western Blot

Laemmli's method [35] was used for electrophoresis of the tested tissue samples (20 µg of protein for MMP-14 self and MMP-15). It was carried out using a 10% SDS-polyacrylamide gel under non-reducing and reducing conditions. The gels were then blotted for 1 h on nitrocellulose membranes (Sigma) at 100 mA. To block the membranes, they were soaked in 5% (*w*/*v*) skimmed milk powder in TBS-T solution (20 mM Tris/HCl buffer, pH 7.4, 150 mM NaCl, 0.05% (*v*/*v*) Tween 20) for 1 h. After this step, the membranes were incubated with a monoclonal antibody directed against human MMP-14 (R&D systems, Minneapolis, MN, USA; cat. no. MAB918) or a monoclonal antibody directed against human MMP-15 (R&D systems, Minneapolis, MN, USA; cat. no. MAB916) in TBS-T containing 1% bovine serum albumin (w/v), overnight at 4 °C. Membranes were then washed several times in TBS-T buffer. Bound antibodies were detected using secondary antibodies conjugated to alkaline phosphatase. For this, the membranes were placed in a solution of secondary antibody for 1 h at room temperature with gentle mixing. After this step, BCIP/NBT reagent (Sigma-Aldrich (Saint Louis, MO, USA; cat. no. B1911) was added. Pre-stained molecular weight markers (BioRad, Hercules, CA, USA) were used to estimate the molecular weight of MMPs. Representative blots that were obtained are shown in the manuscript.

4.5. Actual and Specific Activities of MMPs

The actual and specific activities of MMP-14 and MMP-15 were measured using a black 96-well flat-bottom microplate that was pre-coated with the appropriate MMP-specific antibody; in particular, the same antibodies previously used in the Western blot analysis were used to coat the wells [34].

The immobilization of the metalloproteinase was achieved by adding 100 microliters of respective sample to each well. Next, the microplate with samples was incubated overnight at 4 °C. TBS-T Buffer (50 mM Tris/HCl pH 7.4, 0.9% NaCl, 0.05% Tween 20) was used to wash out any remaining proteins not bound to the microplate. The activity of the tested MMPs was determined in 100 µL of 50 mM Tris/HCl buffer at pH 7.5 (10 mM $CaCl_2$, 150 mM NaCl, and 0.025% Brij 35) [9] and using MCA-Pro-Leu-Ala-Cys(p-OMeBz)-Trp-Ala-Arg(Dpa)-H2 (Merck, Darmstadt, Germany; cat. no. 444258) as a fluorogenic substrate (4 µM final concentration). For this purpose, the microplate was incubated for 60 min at 37 °C with gentle shaking. After incubation, the reaction was stopped by adding 25 µL of 100 mM $EDTANa_2$. Using a spectrophotofluorimetric/multimode microplate reader (Tecan Infinite® 200 PRO, Männedorf, Switzerland; excitation wavelength 325 nm, emission wavelength 393 nm), the degradation of the fluorogenic substrate was determined. Then, using a calibration curve prepared under the same conditions with 7-amino-4-methylcoumarin (Sigma-Aldrich, Saint Louis, MO, USA; cat. no. 257370) as a standard, the amount of degraded substrate was measured.

The actual activity of each tested MMP was calculated as enzyme activity per kg of total protein content (catal/kg of total protein content). The specific activity, on the other hand, was calculated as the enzyme activity per kg of MMP content in the sample (catal/kg of MMP content).

4.6. Protein Determination

The Bradford protein assay was used to measure the protein concentration [36].

4.7. Statistical Analysis

Mean values were calculated based on 10 obtained results for each tested group. Their standard deviations (SD) were also determined. Statistical analysis was carried out using Student's t-test, taking $p < 0.05$ as indicating statistical significance.

5. Conclusions

Urinary tract cancers constitute an important diagnostic and therapeutic problem, and their incidence increases every year. Renal cell carcinoma is the most common type of malignant tumor in the kidney. Extracellular matrix metalloproteinases play an important role in cancer processes due to their involvement in local tumor invasion, the formation of distant metastases, and the process of angiogenesis. The aim of this study was to evaluate the expression, content, and activity of selected membrane-type metalloproteinases (MMP-14 and MMP-15) in human renal cell carcinoma. The research and analysis carried out showed that both control and cancerous human kidney tissues contain MMP-14 and MMP-15 enzymes in high-molecular complexes, in both active and latent forms. Their content is similar in cancer tissues but decreases with higher kidney cancer grades. Notably, MMP-14 is most abundant in control tissues and dominates over MMP-15 in all tissues. The activity of both enzymes increases with cancer grade. MMP-14 appears to play a more significant role than MMP-15 in renal carcinoma, suggesting it could be a promising target for diagnosis, prognosis, and therapy in renal cell carcinoma.

Author Contributions: Conceptualization, G.M., A.T., M.G.-S., B.M., P.M., V.N., A.N. and L.R.; methodology, A.T., G.M., M.G.-S., B.M., P.M., V.N., A.N. and L.R.; validation, G.M., A.T. and L.R.; formal analysis, G.M., A.T. and L.R.; investigation, G.M., A.T. and L.R.; resources, G.M. and L.R.; data curation, G.M., A.T. and L.R.; writing—original draft preparation, G.M., A.T., M.G.-S., P.M., V.N., A.N. and L.R.; writing—review and editing, G.M., A.T., M.G.-S., B.M., V.N. and L.R.; visualization, A.T., G.M. and L.R.; supervision, G.M., A.T. and L.R.; project administration, A.T., G.M. and L.R.; funding acquisition, G.M. and L.R. All authors have read and agreed to the published version of the manuscript.

Funding: This research was funded by the Medical University of Białystok, Poland.

Institutional Review Board Statement: The study was conducted in accordance with the Declaration of Helsinki and approved by the Bioethical Committee of the Medical University of Bialystok (R-I-002/220/2015, 28 May 2015).

Informed Consent Statement: Informed consent was obtained from all subjects involved in the study.

Data Availability Statement: All data generated or analyzed during this study are included in this article. Further enquiries can be directed to the corresponding author.

Acknowledgments: The authors wish to thank Barbara Darewicz, from the Department of Urology, for the possibility of collecting the material of patients with bladder cancer. The authors would also like to thank Tomasz Gogiel, from the Department of Medical Biochemistry, for his valuable comments and assistance during the statistical analysis.

Conflicts of Interest: The authors declare no conflicts of interest.

References

1. Netter, F. *Atlas of Human Anatomy*, 6th ed.; Elsevier: Amsterdam, The Netherlands, 2014.
2. Bochenek, A.; Reicher, M. *Anatomia Człowieka Tom II*; PZWL: Warsaw, Poland, 2004; pp. 476–537.
3. Górski, J. *Fizjologiczne Podstawy Wysiłku Fizycznego*; PZWL: Warsaw, Poland, 2008; pp. 368–387.
4. Boor, P.; Sebekova, K.; Ostendorf, T.; Floege, J. Treatment targets in renal fibrosis. *Nephrol. Dial. Transpl.* **2007**, *22*, 3391–3407. [CrossRef]
5. Catania, J.M.; Chen, G.; Parrish, A.R. Role of matrix metalloproteinases in renal pathophysiologies. *Am. J. Physiol. Renal. Physiol.* **2007**, *292*, F905–F911. [CrossRef]
6. Pardo, A.; Selman, M. Matrix metalloproteases in aberrant fibrotic tissue remodeling. *Proc. Am. Thorac. Soc.* **2006**, *3*, 383–388. [CrossRef]
7. Castro-Castro, A.; Marchesin, V.; Monteiro, P.; Lodillinsky, C.; Rossé, C.; Chavrier, P. Cellular and molecular mechanisms of MT1-MMP-dependent cancer cell Invasion. *Annu. Rev. Cell Dev. Biol.* **2016**, *32*, 555–576. [CrossRef]
8. Pittayapruek, P.; Meephansan, J.; Prapapan, O.; Komine, M.; Ohtsuki, M. Role of Matrix Metalloproteinases in Photoaging and Photocarcinogenesis. *Int. J. Mol. Sci.* **2016**, *17*, 868. [CrossRef]
9. Watkins, G.A.; Fung Jones, E.; Scott Shell, M.; VanBrocklin, H.F.; Pan, M.-H.; Hanrahan, S.M.; Feng, J.J.; He, J.; Sounni, N.E.; Dill, K.A.; et al. Development of an optimized activatable MMP-14 targeted SPECT imaging probe. *Bioorg. Med. Chem.* **2009**, *17*, 653–659. [CrossRef]
10. Amar, S.; Smith, L.; Fields, G.B. Matrix metalloproteinase collagenolysis in health and disease. *Biochim. Biophys. Acta Mol. Cell Res.* **2017**, *1864*, 1940–1951. [CrossRef]
11. Yoshizaki, T.; Sato, H.; Furukawa, M. Recent advances in the regulation of matrix metalloproteinase 2 activation: From basic research to clinical implication (Review). *Oncol. Rep.* **2002**, *9*, 607–611. [CrossRef]
12. Ingvarsen, S.; Madsen, D.H.; Hillig, T.; Lund, L.R.; Holmbeck, K.; Behrendt, N.; Engelholm, L.H. Dimerization of endogenous MT1-MMP is a regulatory step in the activation of the 72-kDa gelatinase MMP-2 on fibroblasts and fibrosarcoma cells. *Biol. Chem.* **2008**, *389*, 943–953. [CrossRef]
13. Itoh, Y.; Ito, N.; Nagase, H.; Evans, R.D.; Bird, S.A.; Seiki, M. Cell surface collagenolysis requires homodimerization of the membrane-bound collagenase MT1-MMP. *Mol. Biol. Cell.* **2006**, *17*, 5390–5399. [CrossRef]
14. Itoh, Y.; Ito, N.; Nagase, H.; Seiki, M. The second dimer interface of MT1-MMP, the transmembrane domain, is essential for ProMMP-2 activation on the cell surface. *J. Biol. Chem.* **2008**, *283*, 13053–13062. [CrossRef]

15. Sabeh, F.; Ota, I.; Holmbeck, K.; Birkedal-Hansen, H.; Soloway, P.; Balbin, M.; Lopez-Otin, C.; Shapiro, S.; Inada, M.; Weiss, S.J. Tumor cell traffic through the extracellular matrix is controlled by the membrane-anchored collagenase MT1-MMP. *J. Cell Biol.* **2004**, *167*, 769–781. [CrossRef]
16. Szabova, L.; Chrysovergis, K.; Yamada, S.; Holmbeck, K. MT1-MMP is required for efficient tumor dissemination in experimental metastatic disease. *Oncogene* **2008**, *27*, 3274–3281. [CrossRef]
17. Andreucci, M.; Provenzano, M.; Faga, T.; Michael, A.; Patella, G.; Mastroroberto, P.; Serraino, G.F.; Bracale, U.M.; Ielapi, N.; Serra, R. Aortic Aneurysms, Chronic Kidney Disease and Metalloproteinases. *Biomolecules* **2021**, *11*, 194. [CrossRef]
18. Turunen, S.P.; Tatti-Bugaeva, O.; Lehti, K. Membrane-type matrix metalloproteases as diverse effectors of cancer progression. *Biochim. Biophys. Acta* **2017**, *1864 Pt A*, 1974–1988. [CrossRef]
19. Mylona, E.; Nomikos, A.; Magkou, C.; Kamberou, M.; Papassideri, I.; Keramopoulos, A.; Nakopoulou, L. The clinicopathological and prognostic significance of membrane type 1 matrix metalloproteinase (MT1-MMP) and MMP-9 according to their localization in invasive breast carcinoma. *Histopathology* **2007**, *50*, 338–347. [CrossRef]
20. Shanbhogue, A.K.; Prasad, S.R.; Takahashi, N.; Vikram, R.; Sahani, D.V. Recent advances in cytogenetics and molecular biology of adult hepatocellular tumors: Implications for imaging and management. *Radiology* **2011**, *258*, 673–693. [CrossRef]
21. Zhang, L.; Jin, S.; Wei, Y.; Wang, C.; Zou, H.; Hu, J.; Jia, W.; Pang, L. Prognostic Significance of Matrix Metalloproteinase 14 in Patients with Cancer: A Systematic Review and Meta-Analysis. *Clin. Lab.* **2020**, *66*, 745. [CrossRef]
22. Available online: http://www.uniprot.org/uniprot/P45452 (accessed on 23 July 2024).
23. Available online: https://www.wikigenes.org/e/gene/e/4322.html (accessed on 23 July 2024).
24. Available online: http://www.genecards.org/cgi-bin/carddisp.pl (accessed on 23 July 2024).
25. Kitagawa, Y.; Kunimi, K.; Uchibayashi, T.; Sato, H.; Namiki, M. Expression of messenger RNAs for membrane-type 1, 2, and 3 matrix metalloproteinases in human renal cell carcinomas. *J. Urol.* **1999**, *162 Pt 1*, 905–909. [CrossRef]
26. Bosman, F.T.; Stamenkovic, I. Functional structure and composition of the extracellular matrix. *J. Pathol.* **2003**, *200*, 423–428. [CrossRef]
27. Bonnans, C.; Chou, J.; Werb, Z. Remodelling the extracellular matrix in development and disease. *Nat. Rev. Mol. Cell Biol.* **2014**, *15*, 786–801. [CrossRef] [PubMed]
28. Chou, C.J.; Affolter, M.; Kussmann, M. A nutrigenomics view of protein intake: Macronutrient, bioactive peptides, and protein turnover. *Prog. Mol. Biol. Transl. Sci.* **2012**, *108*, 51–74. [CrossRef] [PubMed]
29. Qiao, Z.K.; Li, Y.L.; Lu, H.T.; Wang, K.L.; Xu, W.H. Expression of tissue levels of matrix metalloproteinases and tissue inhibitors of metalloproteinases in renal cell carcinoma. *World J. Surg. Oncol.* **2013**, *11*, 1. [CrossRef]
30. Amin, M.B.; Edge, S.; Greene, F.; Byrd, D.R.; Brookland, R.K.; Washington, M.K.; Gershenwald, J.E.; Compton, C.C.; Hess, K.R.; Sullivan, D.C.; et al. (Eds.) *AJCC Cancer Staging Manual*; Springer Science & Business Media: Berlin/Heidelberg, Germany, 2017.
31. Cannon, G.M., Jr.; Getzenburg, R.H. Urinary matrix metalloproteinases activity is not significantly altered in patients with renal cell carcinoma. *Urology* **2006**, *67*, 848–850. [CrossRef] [PubMed]
32. Genovese, F.; Manresa, A.A.; Leeming, D.J.; Karsdal, M.A.; Boor, P. The extracellular matrix in the kidney: A source of novel non-invasive biomarkers of kidney fibrosis? *Fibrogenesis Tissue Repair* **2014**, *7*, 4. [CrossRef] [PubMed]
33. Lee, D.G.; Yang, K.E.; Hwang, J.W.; Kang, H.S.; Lee, S.Y.; Choi, S.; Shin, J.; Jang, I.S. An, as Indicators of Post-Mortem Interval in a Rat Model, with Use of Lateral Flow Technology. *PLoS ONE* **2016**, *11*, e0160557. [CrossRef]
34. Shi, Z.D.; Ji, X.Y.; Qazi, H.; Tarbell, J.M. Interstitial flow promotes vascular fibroblast myofibroblast smooth muscle cell motility in 3-D collagen I via upregulation of MMP-1. *Am. J. Physiol. Heart Circ. Physiol.* **2009**, *297*, H1225–H1234. [CrossRef]
35. Laemmli, U.K. Cleavage of structural proteins during the assembly of the head of bacteriophage T4. *Nature* **1970**, *227*, 680–685. [CrossRef]
36. Bradford, M.M. A rapid and sensitive method for the quantitation of microgram quantities of protein utilizing the principle of protein-dye binding. *Anal. Biochem.* **1976**, *72*, 248–254. [CrossRef]

International Journal of
Molecular Sciences

MDPI

Case Report

Metastatic Penile Squamous Cell Carcinoma Responsive to Enfortumab Vedotin

Catherine C. Fahey [1,2,†], Caroline A. Nebhan [3,†], Sally York [1,2,4], Nancy B. Davis [1,4], Paula J. Hurley [1,4,5], Jennifer B. Gordetsky [4,5,6] and Kerry R. Schaffer [1,4,*]

1 Department of Medicine, Vanderbilt University Medical Center, Nashville, TN 37232, USA; catherine.fahey@vumc.org (C.C.F.)
2 Tennessee Valley Healthcare System, Veterans' Affairs, Nashville, TN 37232, USA
3 Intermountain Health, Murry, UT 84107, USA
4 Vanderbilt-Ingram Cancer Center, Nashville, TN 37232, USA
5 Department of Urology, Vanderbilt University Medical Center, Nashville, TN 37232, USA
6 Department of Pathology, Microbiology and Immunology, Vanderbilt University Medical Center, Nashville, TN 37232, USA
* Correspondence: kerry.schaffer@vumc.org
† These authors contributed equally to this work.

Abstract: Penile squamous cell carcinoma is a rare disease with very limited data to guide treatment decisions. In particular, there is minimal evidence for effective therapies in the metastatic setting. Here, we present a case of metastatic penile squamous cell carcinoma with response to the Nectin-4 inhibitor enfortumab-vedotin-ejfv (EV). EV was selected due to the evidence of the high expression of Nectin-4 in squamous cell carcinomas, including penile carcinoma. The patient had both radiographic and symptomatic improvement after two cycles of treatment, despite having been treated with multiple prior lines of traditional chemotherapy. This case provides support for the use of antibody–drug conjugates (ADC), including EV, in this disease with few other options in the advanced setting. Further studies examining Nectin-4 and ADCs in penile squamous cell carcinoma should be completed, as high-quality evidence is needed to guide treatment after initial progression for these patients.

Keywords: penile squamous cell carcinoma; Nectin-4; antibody-drug conjugate; targeted therapy; metastatic penile cancer

1. Introduction

Penile squamous cell carcinoma (SCC) is a relatively rare malignancy in the United States, with an expected incidence of 2000 cases and 500 deaths in 2022 [1]. Cases are more common in African and South American countries, where penile carcinoma can represent up to 10% of malignancies in men [2]. A paucity of cases renders therapeutic development in randomized clinical trials for this cancer difficult, and the five-year overall survival for metastatic disease approaches 0% [3,4]. Reflecting this challenge, the current recommended penile cancer-specific therapies in the National Comprehensive Cancer Network (NCCN) guidelines are based on data published 10+ years ago [5]. There are several ongoing clinical trials exploring novel treatment approaches for this rare cancer, including trials investigating single agents and combination therapies of immunotherapy, PARP inhibitors, traditional chemotherapies, VEGF inhibitors, EGFR inhibitors, and engineered T cells [6]. Currently, however, compelling data for recurrent/refractory disease is lacking, and there is a need for additional effective systemic therapies for advanced penile cancer.

Penile SCC can be HPV-mediated or non-HPV mediated, with upwards of 40% of cases showing p16 positivity [7]. Additional risk factors for the development of penile SCC include phimosis, lack of circumcision, tobacco use, poor hygiene, genital warts, urinary tract infections, and obesity [8]. Molecular changes commonly seen in penile carcinogenesis include alterations in RAS and MYC, changes to E-cadherin expression, and changes in the p53 and Rb pathways [2]. Additionally, Nectin-4 has been shown to be highly expressed in squamous cell carcinoma, including penile SCC [9]. Nectin-4 has multiple cellular roles, including as a member of the adherens junction [10] and as a ligand for immune checkpoint inhibition [11]. In a cohort of 57 cases of penile SCC, 89% had >25% of cells show positive staining for Nectin-4, versus no staining in nonmalignant penile tissue. Given this biologic finding, there is a logical rationale for the therapeutic potential of Nectin-4 targeted agents in this rare patient population.

First granted accelerated FDA approval in December 2019, the Nectin-4-directed antibody–drug conjugate (ADC) Padcev™ (enfortumab vedotin-ejfv) (EV) is approved for use in advanced or metastatic urothelial cancer after progression with a PD-1 inhibitor and platinum-based chemotherapy (if eligible) [12]. The ADC binds to Nectin-4 on the cell surface and is internalized, allowing for the proteolytic cleavage of the microtubule-disrupting agent monomethyl auristatin E (MMAE), leading to cell death. Here, we present a case of metastatic penile squamous cell carcinoma with response to EV and discuss the biologic rationale supporting a role for this therapeutic class in this rare cancer.

2. Case Description

A 67-year-old man presented with penile pain and bleeding after a trauma. His medical co-morbidities were notable for hypertension and prior tobacco use. The exam showed a painless, fluctuant nodule, <1 cm in diameter, located at the corona of the glans penis. A percutaneous biopsy of the penile lesion revealed p16-positive, HPV-associated, basaloid squamous cell carcinoma. A pelvic MRI revealed an indeterminate nodular density at the distal penile head and a CT urogram revealed pulmonary lesions concerning for metastatic disease. The patient underwent a partial penectomy, which showed pT3 disease with tumor invasion into the lamina propria, corpus spongiosum, corpus cavernosum, and tunica albuginea. Lymphovascular invasion was identified. The surgical margins were positive; the tumor extended to the skin resection margin. PET-CT imaging revealed multiple bilateral pulmonary nodules with strong F-fluorodeoxyglucose (FDG) avidity, the largest measuring 1.1 cm. Hypermetabolic activity was also observed along the residual distal penile shaft. A bronchoscopy with biopsy confirmed metastatic p16-positive SCC.

After metastatic disease was confirmed, the patient was treated with cisplatin, ifosfamide, and paclitaxel (TIP) systemic chemotherapy for a planned total of four cycles. TIP treatment was complicated by a grade 3 peripheral neuropathy adverse event. CT imaging after the completion of treatment demonstrated a partial response of multiple lung nodules including a decrease of seven nodules and the complete resolution of one. After the planned four cycles, he was monitored with serial CT scans every 3 months. Radiographic progression occurred 13 months after TIP treatment was completed and revealed a new right liver lesion, the increasing size of multiple pulmonary nodules, and a new heterogenous pelvic enhancement involving the prostate and bladder, including the left ureterovesicular junction. In addition to radiographic progression, the patient reported increased pelvic pain and urinary frequency. At that time of progression, the patient was treated with ifosfamide and paclitaxel (IP). His persistent grade 1 neuropathy was felt by his physician team to be a contraindication for cisplatin use, particularly in combination with paclitaxel. He experienced stable disease for four months before developing progression of the pulmonary nodules and pelvic disease.

The patient remained interested in treatment and had a good functional status, with an Eastern Cooperative Oncology Group (ECOG) performance status of 1. Next generation sequencing was completed on the original penectomy specimen and showed no FDA-approved actionable alterations. The following pathogenic alterations were found: FBXW7 S476I, JUN amplification, MAPK1 E322K, MYCL1 amplification, and TSC2 E1583fs*14. The sample was microsatellite stable (MSS) and had a tumor mutation burden (TMB) of 0 Muts/Mb. The provider and patient discussed participation in a clinical trial, but ultimately the patient declined to enroll. Based on the limited standard of care regimens for advanced penile cancer, the off-label use of EV was identified as a potential therapeutic option. This was supported based on the mechanism of EV as an ADC against Nectin-4, which has high expression in squamous cell neoplasms [13,14], a documented response in urothelial carcinomas with squamous features [15], and staining on the penectomy specimen (outlined below).

The patient was treated with EV (1.25 mg/kg) on days 1, 8, and 15 of a 28-day cycle. The patient experienced marked improvement of his pelvic pain within one week of EV initiation, and the complete resolution of pain after completion of cycle 1. Of note, the patient had no other changes in his pain regimen or other analgesic measures such as palliative radiation. During cycle 2, the patient experienced worsening of his peripheral neuropathy from grade 1 to grade 2. Due to this adverse event, he was placed on a 3-week treatment break. The break improved his neuropathy, but delayed cycle 3. The patient also experienced a grade 1 rash, which responded to topical corticosteroids. A CT scan prior to the initiation of cycle 3 showed an interval decrease in the hepatic lesion, the resolution of the index pulmonary nodule, and improvement of the pelvic lesion (Figure 1).

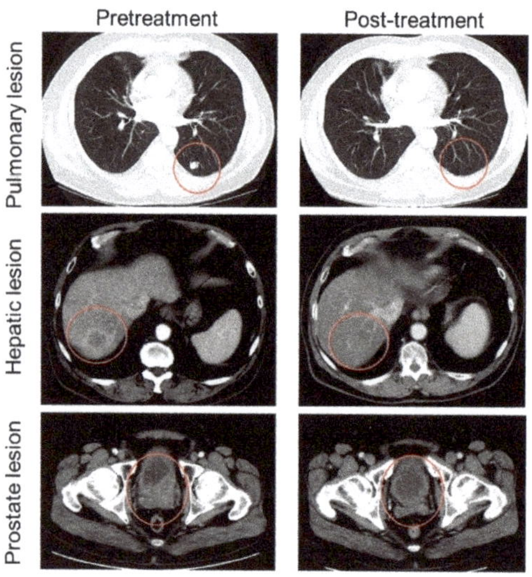

Figure 1. Radiographic response of patient with advanced penile carcinoma before and after treatment with EV. Pre-treatment images (**left**) taken 21 days before EV initiation. Post-treatment images (**right**) were obtained cycle 2 day 27. Resolution of index pulmonary lesion (**top**), near-resolution of R liver lesion (**middle**), and improvement in pelvic metastasis and concurrent posterolateral bladder wall invasion (**bottom**) are indicated by red circles.

Nectin-4 antibody (AB) staining was performed on the penectomy pathology specimen. Slides underwent IHC staining using the Leica Bond-RX and a commercially available anti-Nectin-4 AB (Cat# PA5-50463, ThermoFisher, Carlsbad, CA, USA). Nectin-4 expression was assessed by H-score based on staining intensity (0: none, 1: low, 2: medium, 3: high) and prevalence (0: 0%, 1: 1–33%, 2: 34–66%, 3: 67–100%). The IHC staining showed high Nectin-4 expression: IHC score was intensity (3+) and prevalence (3+) (Figure 2). IHC scoring and slide review was performed by a board-certified pathologist. This work was performed under VUMC IRB 191300.

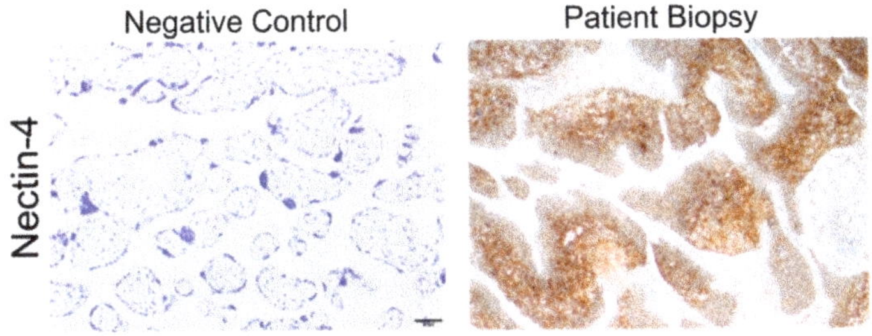

Figure 2. Nectin-4 expression in patient penile tumor. Immunohistochemistry staining of Nectin-4 in negative control (placental tissue, **left panel**) and patient specimen of penile tumor collected during penectomy (**right panel**). Magnification is shown at 20×; bar in left panel represents 50 μm.

The patient had restaging imaging after cycle 4, which showed a mixed response with progression of the lesion in the liver and stable disease in the lung and pelvic lesions. After a discussion of the risks and benefits, and given his ongoing pain control in the pelvis, the decision was made to continue treatment with EV and reassess the response with short interval monitoring. The repeat imaging after cycle 5 of EV showed further progression. The patient was transitioned to pembrolizumab but progressed after three cycles (2 months). Pembrolizumab was selected given its prior use in TMB-high penile cancer [16] and the evidence for its efficacy in other TMB-low tumors [17,18], including SCC [19], though a low likelihood of a response was acknowledged with this regimen. The patient then received gemcitabine [20,21] for two cycles (the patient declined any additional platinum chemotherapy) but the therapy was not tolerated. He then received panitumumab [22] for one cycle but his ECOG performance status declined quickly and he was transitioned to hospice.

3. Discussion

Penile cancer is an aggressive malignancy with poor prognosis and very limited effective treatment options for advanced disease. There is a critical need for novel therapeutic approaches in this cancer. However, the low incidence of cases limits the ability to study treatments in randomized clinical trials. In our case study, we have demonstrated the high Nectin-4 expression of a penile cancer tissue sample paired with the patient's response as preliminary evidence to explore this class of therapy as a novel treatment for penile cancer.

Per NCCN guidelines, the preferred first-line therapy for metastatic penile SCC is paclitaxel/ifosfamide/cisplatin (TIP) [5], as cisplatin-containing regimens have shown the most activity in this disease [23]. 5-fluorouracil/cisplatin can be considered as an alternative first-line chemotherapy [24]. After progression on first-line chemotherapy, patients have overall poor responses. In a 30-patient cohort study of patients with stage TxN2-3M0 penile squamous cell carcinoma treated with TIP prior to lymphadenectomy, 19 patients progressed or recurred [25]. Seventeen of these patients received subsequent systemic therapy, with a median survival from first treatment failure of 5.7 months. Pembrolizumab can be considered in patients with microsatellite instability-high (MSI-high) tumors, as it has been studied in a tumor-agnostic manner and shown to have efficacy [26], [27]. There are case reports of the use of pembrolizumab in penile cancer that support the consideration of this treatment for MSI-high tumors or patients with high PD-L1 expression [16,28]. This is unfortunately a limited population, only estimated to be 15% of patients with advanced penile cancer [29]. There is case report support for combining immunotherapy treatment with radiation [30]. Cetuximab is also identified as a potential therapy based on a small retrospective study, in which 24% of patients had a partial response, and the median time to progression was 2.8 months [31]. There is also evidence that the HER2 pathway may have therapeutic implications for penile SCC [8]. A phase II study of dacomitinib, a pan-HER2 TKI which also inhibits EGFR, was conducted in patients with locally advanced (n = 20) or metastatic (n = 8) penile SCC [32]. In this trial, the overall response rate (ORR) was

32% for all patients, with a median progression-free survival of 3.2 months for patients with metastatic disease. Dacomitinib was shown to be most efficacious for patients with mutations in the downstream effectors of HER receptors or those with TERT mutations. These results, as well as the indications for immunotherapy, support genomic sequencing for patients at the time of progression, as important therapeutic targets may be identified.

Given the lack of robust standard treatment options, NCCN naturally recommends consideration for clinical trials. There are a number of ongoing trials for advanced penile cancer currently exploring classes of therapy including chemotherapy, VEGF/EGFR inhibitors, immunotherapies, and PARP inhibitors [6]. There are several ongoing trials exploring EV in other cancer types at present, some of which include squamous cell histologic subtypes. However, there are no studies evaluating a penile cancer cohort [33].

Nectin-4 is a type I immunoglobulin-like membrane protein found primarily in adherens junctions. It has been shown to affect cell proliferation [34], as well as tumor progression and metastasis [35]. The Nectin-4-directed ADC EV was first granted accelerated FDA approval in December 2019 and is approved for use in advanced or metastatic UC after progression on a PD-1 inhibitor and platinum-based chemotherapy [36]. The human IgG1 ADC binds to Nectin-4 on the cell surface and is internalized, allowing for the proteolytic cleavage of the microtubule-disrupting agent monomethyl auristatin E (MMAE) [12,37,38]. In a phase II trial of EV in UC, the ORR exceeded 50%; this ORR was particularly impressive given the significant pre-treatment with both platinum chemotherapies and PD-1/PD-L1 inhibitors [15]. Of note, in this trial, patients with UC with squamous features on histology were eligible and comprised 13% of the study population. Since its initial approval as monotherapy, EV has also been approved in combination with pembrolizumab for patients with locally advanced or metastatic UC who are ineligible for cisplatin-containing chemotherapy, and demonstrates a significant OS and PFS benefit in previously untreated locally advanced or mUC (including both cisplatin-eligible and -ineligible patients) [39].

EV use in UC does not require Nectin-4 positivity on tissue staining to qualify for treatment. However, numerous basic histopathology studies suggest that this therapy has great potential in other squamous neoplasms. Nectin-4 is highly expressed in squamous cell neoplasms including head and neck [40], lung [41], extramammary Paget's [42], the squamous variant of muscle invasive bladder [43], and penile [9]. A recent study of primary penile cancers by Grass et al. demonstrated high Nectin-4 gene expression in the majority of penile cancer specimens: 70% of the samples had an at least 2-fold higher gene expression of Nectin-4 compared to normal glans [9]. This study also found 61% of the cases had moderate to strong Nectin-4 IHC staining. Importantly, recent preclinical data suggests that EV may have a "bystander" effect, targeting Nectin-4-negative cells in a mixed cellular assay, suggesting a potential role for this drug even in the setting of heterogeneous Nectin-4 expression [44]. There are also now numerous other drugs early in development targeting Nectin-4 including CAR-T therapy and novel ADCs [45].

In our patient, the response time to EV of 5 months was on par with other palliative systemic therapies, despite this patient's substantial prior treatments. He tolerated the therapy well, with neuropathy and rash as the only reported side effects. EV treatment also provided substantial symptomatic improvement for him due to the response of the large pelvic lesion. Overall, the duration of the response and symptomatic improvement with low side effects provide support for EV use as a reasonable therapeutic option for metastatic penile cancer patients that are interested in systemic therapy if clinical trials are not an option. Of course, further studies should be performed to confirm these initial findings.

4. Conclusions

This case report provides promising evidence for the use of EV as a systemic therapy for metastatic penile cancer and reinforces prior work [9] demonstrating Nectin-4 expression in this rare cancer. To our knowledge, this is the first reported case of an ADC with a treatment benefit for penile cancer, including a symptomatic improvement and radiographic response. These findings may be important given the paucity of existing treatment options for this disease, hence, this warrants intensified basic research as well as the realization of clinical therapeutic trials.

Author Contributions: Conceptualization, C.A.N. and K.R.S.; methodology, C.A.N., P.J.H. and J.B.G., formal analysis, J.B.G.; writing—original draft preparation C.C.F., C.A.N., K.R.S.; writing—review and editing C.C.F., P.J.H., J.B.G. and K.R.S.; supervision, S.Y., N.B.D. and K.R.S.; funding acquisition, C.A.N. and K.R.S. All authors have read and agreed to the published version of the manuscript.

Funding: Support for staining was from C.A.N. grant F32CA254070. C.C.F. is supported by T32CA217834. K.R.S. is supported by K12CA090625. The Translational Pathology Shared Resource is supported by NCI/NIH Cancer Center Support Grant P30CA068485, the Shared Instrumentation Grant S10 OD023475-01A1, and the shared equipment grant S10 OD016355.

Institutional Review Board Statement: The study was conducted in accordance with the Declaration of Helsinki, and approved by the Institutional Review Board of Vanderbilt University Medical Center (protocol code 19130).

Informed Consent Statement: Informed consent was obtained from the patient for the publication of his case prior to his passing.

Data Availability Statement: No new data were created or analyzed in this study. Data sharing is not applicable to this article.

Acknowledgments: We acknowledge the Translational Pathology Shared Resource and for the Leica Bond RX, the Tissue MicroArray (TMA) Grandmaster and the LCM: IS1BX003154. We would also like to acknowledge and thank the members of the Vanderbilt SWERV writing core for editing and development of this manuscript. We thank the patient for consenting to the publication of his treatment course.

Conflicts of Interest: Author N.B.D. was employed at VUMC during the time of this case. She is currently employed by Merck & Co., Inc. The remaining authors declare that the research presented in this case was conducted in the absence of any commercial or financial relationships that could be construed as a conflict of interest.

References

1. Siegel, R.L.; Miller, K.D.; Fuchs, H.E.; Jemal, A. Cancer statistics. *CA Cancer J. Clin.* **2022**, *72*, 7–33. [CrossRef] [PubMed]
2. Bleeker, M.C.G.; Heideman, D.A.M.; Snijders, P.J.F.; Horenblas, S.; Dillner, J.; Meijer, C.J.L.M. Penile cancer: Epidemiology, pathogenesis and prevention. *World J. Urol.* **2009**, *27*, 141–150. [CrossRef] [PubMed]
3. Mistretta, F.A.; Palumbo, C.; Knipper, S.; Mazzone, E.; Pecoraro, A.; Tian, Z.; Musi, G.; Perrotte, P.; Montanari, E.; Shariat, S.; et al. Conditional survival of patients with stage I-III squamous cell carcinoma of the penis: Temporal changes in cancer-specific mortality. *World J. Urol.* **2020**, *38*, 725–732. [CrossRef]
4. Horenblas, S. Lymphadenectomy for squamous cell carcinoma of the penis. Part 2: The role and technique of lymph node dissection. *BJU Int.* **2001**, *88*, 473–483. [CrossRef]
5. National Comprehensive Cancer Network® (NCCN®). NCCN Clinical Practice Guidelines in Oncology (NCCN Guidelines®) Penile Cancer Version 1.2023. Available online: https://www.nccn.org/professionals/physician_gls/pdf/penile.pdf (accessed on 1 October 2023).
6. ClinicalTrials.gov. Available online: https://clinicaltrials.gov/search?cond=penile%20cancer&viewType=Table (accessed on 6 July 2023).
7. Eich, M.L.; Pena, M.D.; Schwartz, L.; Granada, C.P.; Rais-Bahrami, S.; Giannico, G.; Amador, B.M.; Matoso, A.; Gordetsky, J.B. Morphology, p16, HPV, and outcomes in squamous cell carcinoma of the penis: A multi-institutional study. *Hum. Pathol.* **2019**, *96*, 79–86. [CrossRef] [PubMed]
8. Chadha, J.; Chahoud, J.; Spiess, P.E. An update on treatment of penile cancer. *Ther. Adv. Med. Oncol.* **2022**, *14*, 17588359221127254. [CrossRef] [PubMed]
9. Grass, G.D.; Chahoud, J.; Lopez, A.; Dhillon, J.; Eschrich, S.A.; Johnstone, P.A.; Spiess, P.E. An Analysis of Nectin-4 (PVRL4) in Penile Squamous Cell Carcinoma. *Eur. Urol. Open Sci.* **2023**, *49*, 1–5. [CrossRef]
10. Liu, Y.; Han, X.; Li, L.; Zhang, Y.; Huang, X.; Li, G.; Xu, C.; Yin, M.; Zhou, P.; Shi, F.; et al. Role of Nectin-4 protein in cancer (Review). *Int. J. Oncol.* **2021**, *59*, 93. [CrossRef]

11. Reches, A.; Ophir, Y.; Stein, N.; Kol, I.; Isaacson, B.; Amikam, Y.C.; Elnekave, A.; Tsukerman, P.; Brlic, P.K.; Lenac, T.; et al. Nectin4 is a novel TIGIT ligand which combines checkpoint inhibition and tumor specificity. *J. Immunother. Cancer* **2020**, *8*, e000266. [CrossRef]
12. Alhalabi, O.; Altameemi, L.; Campbell, M.T.; Meric-Bernstam, F. Recent Advances in the Development of Antibody-Drug Conjugates in Urothelial Cancer. *Cancer J.* **2022**, *28*, 417. [CrossRef]
13. Tanaka, Y.; Murata, M.; Oda, Y.; Furue, M.; Ito, T. Nectin Cell Adhesion Molecule 4 (NECTIN4) Expression in Cutaneous Squamous Cell Carcinoma: A New Therapeutic Target? *Biomedicines* **2021**, *9*, 4. [CrossRef]
14. Hashimoto, H.; Tanaka, Y.; Murata, M.; Ito, T. Nectin-4: A Novel Therapeutic Target for Skin Cancers. *Curr. Treat. Options Oncol.* **2022**, *23*, 578–593. [CrossRef]
15. Yu, E.Y.; Petrylak, D.P.; O'Donnell, P.H.; Lee, J.L.; van der Heijden, M.S.; Loriot, Y.; Stein, M.N.; Necchi, A.; Kojima, T.; Harrison, M.R.; et al. Enfortumab vedotin after PD-1 or PD-L1 inhibitors in cisplatin-ineligible patients with advanced urothelial carcinoma (EV-201): A multicentre, single-arm, phase 2 trial. *Lancet Oncol.* **2021**, *22*, 872–882. [CrossRef] [PubMed]
16. Hahn, A.W.; Chahoud, J.; Campbell, M.T.; Karp, D.D.; Wang, J.; Stephen, B.; Tu, S.M.; Pettaway, C.A.; Naing, A. Pembrolizumab for advanced penile cancer: A case series from a phase II basket trial. *Investig. New Drugs* **2021**, *39*, 1405–1410. [CrossRef] [PubMed]
17. Yarchoan, M.; Hopkins, A.; Jaffee, E.M. Tumor Mutational Burden and Response Rate to PD-1 Inhibition. *N. Engl. J. Med.* **2017**, *377*, 2500–2501. [CrossRef] [PubMed]
18. Goodman, A.M.; Kato, S.; Bazhenova, L.; Patel, S.P.; Frampton, G.M.; Miller, V.; Stephens, P.J.; Daniels, G.A.; Kurzrock, R. Tumor Mutational Burden as an Independent Predictor of Response to Immunotherapy in Diverse Cancers. *Mol. Cancer Ther.* **2017**, *16*, 2598–2608. [CrossRef]
19. Goodman, A.M.; Kato, S.; Chattopadhyay, R.; Okamura, R.; Saunders, I.M.; Montesion, M.; Frampton, G.M.; Miller, V.A.; Daniels, G.A.; Kurzrock, R. Phenotypic and Genomic Determinants of Immunotherapy Response Associated with Squamousness. *Cancer Immunol. Res.* **2019**, *1*, 866–873. [CrossRef]
20. Houédé, N.; Dupuy, L.; Fléchon, A.; Beuzeboc, P.; Gravis, G.; Laguerre, B.; Théodore, C.; Culine, S.; Filleron, T.; Chevreau, C. Intermediate analysis of a phase II trial assessing gemcitabine and cisplatin in locoregional or metastatic penile squamous cell carcinoma. *BJU Int.* **2016**, *117*, 444–449. [CrossRef]
21. Power, D.G.; Galvin, D.J.; Cuffe, S.; McVey, G.P.; Mulholland, P.J.; Farrelly, C.; Delaney, D.W.; O'Byrne, K.J. Cisplatin and gemcitabine in the management of metastatic penile cancer. *Urol. Oncol.* **2009**, *27*, 187–190. [CrossRef]
22. Necchi, A.; Giannatempo, P.; Vullo, S.L.; Raggi, D.; Nicolai, N.; Colecchia, M.; Perrone, F.; Mariani, L.; Salvioni, R. Panitumumab Treatment for Advanced Penile Squamous Cell Carcinoma When Surgery and Chemotherapy Have Failed. *Clin. Genitourin. Cancer* **2016**, *14*, 231–236. [CrossRef]
23. Pettaway, C.A.; Pagliaro, L.; Theodore, C.; Haas, G. Treatment of visceral, unresectable, or bulky/unresectable regional metastases of penile cancer. *Urology* **2010**, *76* (Suppl. S1), S58–S65. [CrossRef] [PubMed]
24. Di Lorenzo, G.; Buonerba, C.; Federico, P.; Perdonà, S.; Aieta, M.; Rescigno, P.; D'Aniello, C.; Puglia, L.; Petremolo, A.; Ferro, M.; et al. Cisplatin and 5-fluorouracil in inoperable, stage IV squamous cell carcinoma of the penis. *BJU Int.* **2012**, *110*, E661–E666. [CrossRef] [PubMed]
25. Wang, J.; Pettaway, C.A.; Pagliaro, L.C. Treatment for Metastatic Penile Cancer After First-line Chemotherapy Failure: Analysis of Response and Survival Outcomes. *Urology* **2015**, *85*, 1104–1110. [CrossRef]
26. Marabelle, A.; Le, D.T.; Ascierto, P.A.; Di Giacomo, A.M.; De Jesus-Acosta, A.; Delord, J.P.; Geva, R.; Gottfried, M.; Penel, N.; Hansen, A.R.; et al. Efficacy of Pembrolizumab in Patients With Noncolorectal High Microsatellite Instability/Mismatch Repair-Deficient Cancer: Results From the Phase II KEYNOTE-158 Study. *J. Clin. Oncol. Off. J. Am. Soc. Clin. Oncol.* **2020**, *38*, 1–10. [CrossRef] [PubMed]
27. Le, D.T.; Durham, J.N.; Smith, K.N.; Wang, H.; Bartlett, B.R.; Aulakh, L.K.; Lu, S.; Kemberling, H.; Wilt, C.; Luber, B.S.; et al. Mismatch repair deficiency predicts response of solid tumors to PD-1 blockade. *Science* **2017**, *357*, 409–413. [CrossRef] [PubMed]
28. Chahoud, J.; Skelton IV, W.P.; Spiess, P.E.; Walko, C.; Dhillon, J.; Gage, K.L.; Johnstone, P.A.; Jain, R.K. Case Report: Two Cases of Chemotherapy Refractory Metastatic Penile Squamous Cell Carcinoma With Extreme Durable Response to Pembrolizumab. *Front. Oncol.* **2020**, *10*, 615298. [CrossRef]
29. Spiess, P.E.; Li, R.; Grivas, P.; Necchi, A.; Pavlick, D.C.; Huang, R.S.; Lin, D.I.; Danziger, N.; Ross, J.S.; Jacob, J.M. Bratslavsky, G. Penile squamous cell carcinoma (PSCC) with elevated tumor mutational burden (TMB): A genomic landscape study. *J. Clin. Oncol.* **2023**, *41* (Suppl. S6), 4. [CrossRef]
30. Kaakour, D.; Seyedin, S.; Houshyar, R.; Mar, N. Combination of Pembrolizumab and Stereotactic Body Radiation Therapy in Recurrent Metastatic Penile Squamous Cell Carcinoma: A Case Study. *Biomedicines* **2022**, *10*, 3033. [CrossRef]
31. Carthon, B.C.; Ng, C.S.; Pettaway, C.A.; Pagliaro, L.C. Epidermal Growth Factor Receptor–Targeted Therapy in Locally Advanced or Metastatic Squamous Cell Carcinoma of the Penis. *BJU Int.* **2014**, *113*, 871–877. [CrossRef]
32. Necchi, A.; Lo Vullo, S.; Perrone, F.; Raggi, D.; Giannatempo, P.; Calareso, G.; Nicolai, N.; Piva, L.; Biasoni, D.; Catanzaro, M.; et al. First-line therapy with dacomitinib, an orally available pan-HER tyrosine kinase inhibitor, for locally advanced or metastatic penile squamous cell carcinoma: Results of an open-label, single-arm, single-centre, phase 2 study. *BJU Int.* **2018**, *121*, 348–356. [CrossRef]

33. Astellas Pharma Global Development, Inc. An Open-label, Multicenter, Multicohort, Phase 2 Study to Evaluate Enfortumab Vedotin in Subjects with Previously Treated Locally Advanced or Metastatic Malignant Solid Tumors (EV-202). clinicaltrials.gov, Clinical trial registration NCT04225117, April 2023. Available online: https://clinicaltrials.gov/ct2/show/NCT04225117 (accessed on 24 May 2023).
34. Nishiwada, S.; Sho, M.; Yasuda, S.; Shimada, K.; Yamato, I.; Akahori, T.; Kinoshita, S.; Nagai, M.; Konishi, N.; Nakajima, Y. Nectin-4 expression contributes to tumor proliferation, angiogenesis and patient prognosis in human pancreatic cancer. *J. Exp. Clin. Cancer Res.* **2015**, *34*, 30. [CrossRef] [PubMed]
35. Sethy, C.; Goutam, K.; Nayak, D.; Pradhan, R.; Molla, S.; Chatterjee, S.; Rout, N.; Wyatt, M.D.; Narayan, S.; Kundu, C.N. Clinical significance of a pvrl 4 encoded gene Nectin-4 in metastasis and angiogenesis for tumor relapse. *J. Cancer Res. Clin. Oncol.* **2020**, *146*, 245–259. [CrossRef] [PubMed]
36. Alt, M.; Stecca, C.; Tobin, S.; Jiang, D.M.; Sridhar, S.S. Enfortumab Vedotin in urothelial cancer. *Ther. Adv. Urol.* **2020**, *12*, 1756287220980192. [CrossRef] [PubMed]
37. Challita-Eid, P.M.; Satpayev, D.; Yang, P.; An, Z.; Morrison, K.; Shostake, Y.; Raitano, A.; Nadell, R.; Liu, W.; Lortie, D.R.; et al. Enfortumab Vedotin Antibody–Drug Conjugate Targeting Nectin-4 Is a Highly Potent Therapeutic Agent in Multiple Preclinical Cancer Models. *Cancer Res.* **2016**, *76*, 3003–3013. [CrossRef]
38. Doronina, S.O.; Toki, B.E.; Torgov, M.Y.; Mendelsohn, B.A.; Cerveny, C.G.; Chace, D.F.; DeBlanc, R.L.; Gearing, R.P.; Bovee, T.D.; Siegall, C.B.; et al. Development of potent monoclonal antibody auristatin conjugates for cancer therapy. *Nat. Biotechnol.* **2003**, *21*, 7. [CrossRef]
39. Thomas, P. ESMO 2023: EV-302/KEYNOTE-A39: Enfortumab Vedotin in Combination with Pembrolizumab (EV+P) Vs Chemotherapy in Previously Untreated Locally Advanced Metastatic Urothelial Carcinoma. Presented at the ESMO Congress 2023, Madrid, Spain, 22 October 2023.
40. Sanders, C.; Lau, J.-F.; Dietrich, D.; Strieth, S.; Brossart, P.; Kristiansen, G. Nectin-4 is widely expressed in head and neck squamous cell carcinoma. *Oncotarget* **2022**, *13*, 1166–1173. [CrossRef]
41. Takano, A.; Ishikawa, N.; Nishino, R.; Masuda, K.; Yasui, W.; Inai, K.; Nishimura, H.; Ito, H.; Nakayama, H.; Miyagi, Y.; et al. Identification of nectin-4 oncoprotein as a diagnostic and therapeutic target for lung cancer. *Cancer Res.* **2009**, *69*, 6694–6703. [CrossRef]
42. Murata, M.; Ito, T.; Tanaka, Y.; Kaku-Ito, Y.; Furue, M. NECTIN4 Expression in Extramammary Paget's Disease: Implication of a New Therapeutic Target. *Int. J. Mol. Sci.* **2020**, *21*, 5891. [CrossRef]
43. Hoffman-Censits, J.H.; Lombardo, K.A.; Parimi, V.; Kamanda, S.; Choi, W.; Hahn, N.M.; McConkey, D.J.; McGuire, B.M.; Bivalacqua, T.J.; Kates, M.; et al. Expression of Nectin-4 in Bladder Urothelial Carcinoma, in Morphologic Variants, and Nonurothelial Histotypes. *Appl. Immunohistochem. Mol. Morphol. AIMM* **2021**, *29*, 619–625. [CrossRef]
44. Liu, B.A.; Olson, D.; Snead, K.; Gosink, J.; Tenn, E.M.; Zaval, M.; Cao, A.; Sahetya, D.; Nesterova, A.; Hensley, K.; et al. Abstract 5581: Enfortumab vedotin, an anti-Nectin-4 ADC demonstrates bystander cell killing and immunogenic cell death anti-tumor activity mechanisms of action in urothelial cancers. *Cancer Res.* **2020**, *80* (Suppl. S16), 5581. [CrossRef]
45. Clinicaltrials.gov. Phase 1/2 Study of the Safety, Pharmacokinetics, and Preliminary Clinical Activity of BT7480 in Patients with Nectin-4 Associated Advanced Malignancies. clinicaltrials.gov, Clinical trial registration NCT05163041, March 2023. Available online: https://clinicaltrials.gov/study/NCT05163041 (accessed on 31 December 2022).

International Journal of *Molecular Sciences*

MDPI

Article

Cellular and Molecular Evidence of the Synergistic Antitumour Effects of Hydroxytyrosol and Metformin in Prostate Cancer

Francisco Porcel-Pastrana [1,2,3], Antonio J. Montero-Hidalgo [1,2,3], Miguel E. G-García [1,2,3], Ignacio Gil-Duque [1,2,3], Antonio Prats-Escribano [1,2,3], Manuel D. Gahete [1,2,3,4], André Sarmento-Cabral [1,2,3], Raúl M. Luque [1,2,3,4,*] and Antonio J. León-González [1,2,3,5,*]

1 Maimonides Institute for Biomedical Research of Córdoba (IMIBIC), 14004 Córdoba, Spain; porcelpastrana.francisco@gmail.com (F.P.-P.); antonio.montero.96@gmail.com (A.J.M.-H.); miedu1998gargar@gmail.com (M.E.G.-G.); igilduque@gmail.com (I.G.-D.); antoniopratsescribano@gmail.com (A.P.-E.); bc2gaorm@uco.es (M.D.G.); amsbcabral@gmail.com (A.S.-C.)
2 Department of Cell Biology, Physiology and Immunology, University of Córdoba, 14014 Córdoba, Spain
3 Hospital Universitario Reina Sofía (HURS), 14004 Córdoba, Spain
4 Centro de Investigación Biomédica en Red de Fisiopatología de la Obesidad y Nutrición (CIBERobn), 28019 Madrid, Spain
5 Department of Pharmacology, School of Pharmacy, University of Seville, 41012 Seville, Spain
* Correspondence: raul.luque@uco.es (R.M.L.); ajleon@us.es (A.J.L.-G.); Tel.: +34-95542-0241 (A.J.L.-G.)

Abstract: Prostate cancer (PCa) is the tumour pathology with the second highest incidence among men worldwide. PCa is strongly influenced by obesity (OB), which increases its aggressiveness. Hence, some metabolic drugs like metformin have emerged as potential anti-tumour agents against several endocrine-related cancers. Likewise, a high adherence to the Mediterranean diet has been associated with lower rates of OB and a reduction in PCa aggressiveness since this diet contains phenolic bioactive compounds such as hydroxytyrosol (HT) that is mainly present in extra virgin olive oil. Thus, we decided to analyse the therapeutic potential of the combination of HT + metformin in different PCa cell models. Specifically, combinations of different doses of HT and metformin were evaluated by analysing the proliferation rate of LNCaP, 22Rv1, DU-145, and PC−3 cells using the SynergicFinder method. The results revealed a synergistic effect of HT + metformin in significantly reducing proliferation, especially in LNCaP cells. This anti-tumour effect of HT + metformin was also confirmed in migration and tumoursphere formation assays in LNCaP. The effects on the cell cycle and apoptosis were also assessed by flow-cytometry, and a cycle arrest in the G1 phase and an increase in late apoptosis were observed with the combination of HT + metformin. The phosphorylation levels of critical components of different oncogenic pathways were measured which revealed that the combination of HT + metformin significantly reduced the activity of multiple components of the MAPK, AKT, and TGF-β pathways. Overall, the combination of HT + metformin might represent a new therapeutic avenue for the management of PCa patients, an observation that certainly warrants further investigation through a well-designed clinical trial.

Keywords: prostate cancer; metformin; hydroxytyrosol; anti-tumour actions

1. Introduction

Prostate cancer (PCa) constitutes the second most frequently diagnosed malignancy and the fifth leading cause of cancer-related mortality among males globally [1]. While the initial diagnosis often identifies tumours with low risk of progression, it is concerning that 30–50% of these cases may progress to advanced PCa [especially to the phenotype of Castration-Resistant Prostate Cancer (CRPC)] within a 5–10-year period, which remains an incurable and ultimately fatal illness [2]. Therefore, patients who develop this advanced PCa require a transition to active therapeutic interventions [3]. Due to the typically indolent progression of PCa, there is a significant interest in identifying therapeutic options with minimal adverse effects and ideally, treatments that are efficacious in all phenotypes of PCa. An emerging approach to treat several cancer types is based on the repositioning of drugs that are already approved to treat other pathologies since this strategy is faster and easier to translate to clinical practice compared to the development and validation of new drugs [4]. In this context, one of the most promising research lines in cancer research focuses on drugs commonly used for treating certain metabolic disorders (e.g., type 2 diabetes, obesity, and hypercholesterolemia), such as metformin or statins (e.g., simvastatin) [5–13]. In fact, several epidemiological studies have reported that patients treated with metformin showed a lower risk of developing various malignancies, including PCa [13–16], and that metformin in combination with androgen deprivation therapy (ADT) could decrease some of the unfavourable metabolic consequences of ADT and enhance the tumour-suppressive effect of ADT [17,18].

Within this framework, numerous investigations have also explored the potential of bioactive compounds derived from edible fruits and vegetables within the Mediterranean diet for the treatment and prevention of PCa [19,20]. Notably, there is a burgeoning interest in bioactive compounds present in extra virgin olive oil (EVOO), particularly phenolic alcohols [e.g., tyrosol and hydroxytyrosol (HT)] and their derivatives (e.g., oleocanthal and other secoiridoids), in many oncological contexts [21]. Among these bioactive compounds, HT has demonstrated promising anticancer properties on different in vitro and in vivo cancer models primarily due to its function as an antioxidant [13,22–27]. In fact, our group previously demonstrated the anti-tumour effects of HT and its chemical derivatives on PCa cells [28].

Interestingly, independent researchers have pinpointed synergic anti-tumour effects of the combination of different bioactive antioxidant compounds with metformin in various cancer types [29]. However, to the best of our knowledge, the potential anti-tumour effects of the combination of HT and metformin in PCa cells remains largely unknown. For that reason, this study was conceived to pursue the following objective: to explore the in vitro anti-tumour effects of the combination of metformin and HT in different human PCa cell models [androgen-sensitive (LNCaP and 22Rv1), androgen-independent (DU-145), and CRPC (PC-3) models] to analyse the ability of these drugs and their combination to alter the activity of multiple key oncogenic signalling pathways that are critical for tumour development, progression, and aggressiveness in human PCa.

2. Results

2.1. Synergic Effect of Hydroxytyrosol and Metformin in Prostate Cancer Cell Models

To determine the potential interaction effect of the combined treatment with HT and metformin, we performed a dose–response matrix with five different concentrations (see Table 1) based on previous results from our group [11,16,28]. To this end, we focused on the inhibition of cell proliferation as a functional readout to identify interaction effects for every combination. We selected LNCaP and 22Rv1 cells as androgen-sensitive cell models, DU-145 cells as an androgen-independent cell model, and PC−3 cells as a CRPC cell model. Based on the IC50 analyses, LNCaP cells were the most sensitive to both treatments with the highest Combinatory Synergy Score (based on SynergyFinder software 2.0 [30]), followed by 22Rv1 cells (both the LNCaP and 22Rv1 cell models showed synergy effects), while

DU-145 and PC−3 cells had the lowest Synergy Scores, showing an additive effect (Table 2). It should be mentioned that the IC50 could not be obtained for metformin or HT in PC−3 cells since a proliferative effect lower than 50% was obtained in response to all the doses used (Table 2).

Table 1. SynergyFinder 2.0 dose design. Doses were selected based on Refs. [8,11,16,28].

SynergyFinder 2.0 Design					
Drug	**Dose 1**	**Dose 2**	**Dose 3**	**Dose 4**	**Dose 5**
Metformin (mM)	0	0.5	1	2.5	5
Hydroxytyrosol (μM)	0	3	10	30	100

Table 2. IC50 values for metformin and hydroxytyrosol treatments and Combinatory Synergy Index Scores (based on SynergyFinder software [30]).

Individual Treatment IC_{50}				
Drug	**LNCaP**	**22Rv1**	**DU-145**	**PC-3**
Metformin (mM)	2.313	4.394	5	-
Hydroxytyrosol (μM)	43.66	74.84	86.30	-
Combinatory Synergy Index	52.08	31.16	6.08	1.379
Combination effect	Synergy	Synergy	Additive	Additive

More specifically, LNCaP cells were the cell model showing the most effective actions in response to the individual or combination treatments (HT and/or metformin; Figure 1A; two-way ANOVA statistics shown in Table 3; Dunnett's multiple comparison test results shown in Appendix A−Table A1). As mentioned above, LNCaP cells also showed a stronger synergy effect (calculated using the Synergy Combination Index; Table 2). For the 22Rv1 cell model, the synergic effect was also evident but was milder than that observed in the LNCaP model (Table 2), with the main effect induced by HT (Figure 1B; two-way ANOVA statistics shown in Table 3; Dunnett's multiple comparison test results shown in Appendix A−Table A1). The proliferation rate was reduced in DU-145 cells only with the maximum doses of both treatments (HT and metformin), but no overall significant interaction was found between them (Figure 1C; two-way ANOVA statistics shown in Table 3; Dunnett's multiple comparison test results shown in Appendix A−Table A1). Finally, the proliferation rate in PC−3 cells was slightly, but significantly, reduced in response to HT or metformin using different doses and showed a low inhibitory interaction with the maximum doses of both treatments (Figure 1D; two-way ANOVA statistics shown in Table 3; Dunnett's multiple comparison test results shown in Appendix A−Table A1). Based on these results, we selected LNCaP cells (an androgen-sensitive model) for the following analyses since it showed the highest sensitivity to the combination of HT and metformin (Tables 2 and 3, and Figure 1). From all the combinations, 10 μM HT and 2.5 mM metformin were selected to further explore the synergic effects of these treatments because this was the combination that was able to exert a prominent synergic effect on the proliferation rate and it reduced the needed dose for both compounds when used individually.

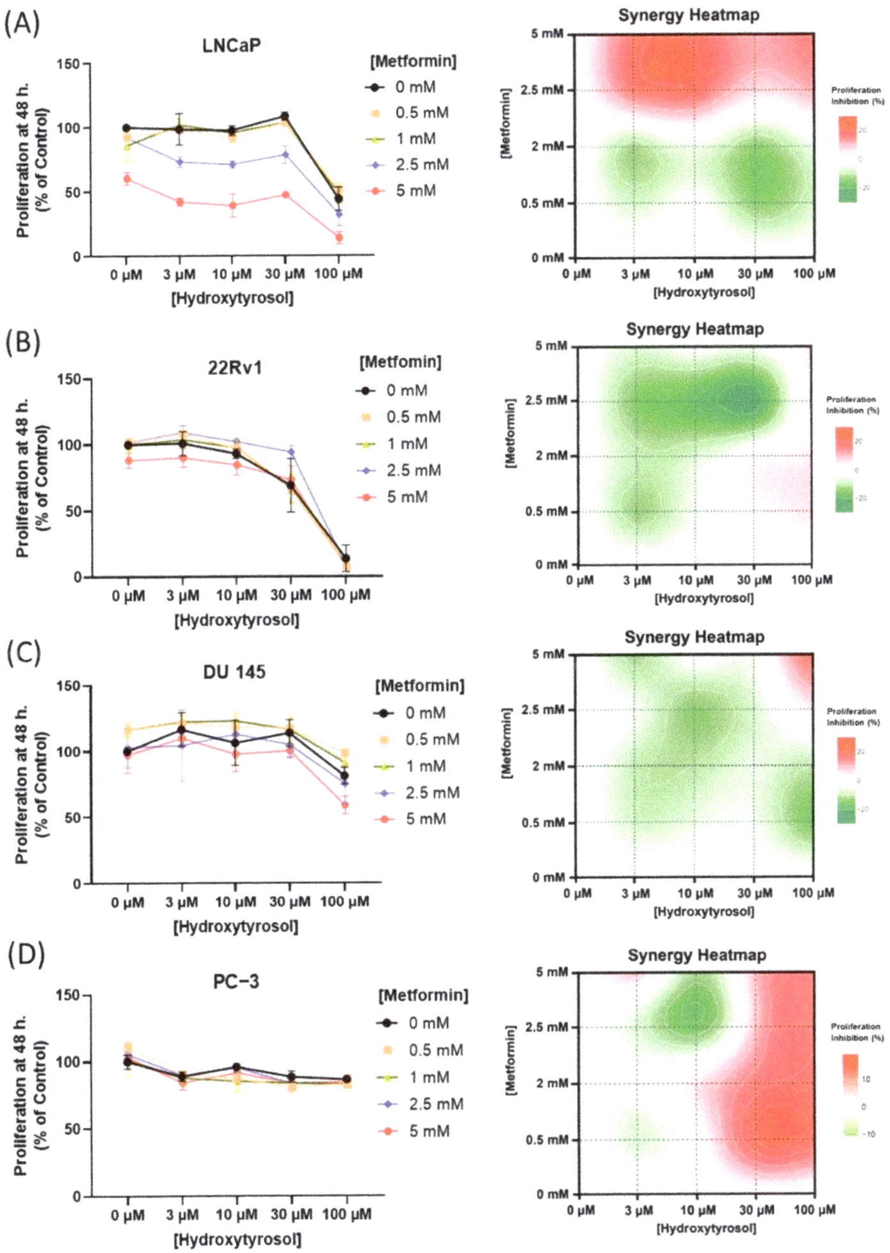

Figure 1. Dose–response curves of hydroxytyrosol and metformin in prostate cancer cell lines. Proliferation rate in response to hydroxytyrosol, metformin, and their different combinations in LNCaP (**A**) left panel), 22Rv1 (**B**) left panel), DU 145 (**C**) left panel), and PC−3 (**D**) left panel) cell lines after 48 h of incubation. Control set at 100%. Data are represented as percentage relative to negative control (mean ± SEM; $n \geq 3$). Heatmaps of synergic scores generated with SynergyFinder 2.0 for LNCaP ((**A**) right panel), 22Rv1 ((**B**) right panel), DU 145 ((**C**) right panel), and PC−3 ((**D**) right panel) cell lines with proliferation inhibition data.

Table 3. Two-way ANOVA statistics for drug interaction effects in cell proliferation assay. Asterisks represent statistical differences based on ANOVA among all doses used for each treatment group and the interaction between treatments (ns, $p > 0.05$; *, $p < 0.05$; ****, $p < 0.0001$).

	LNCaP		22Rv1		DU 145		PC-3	
Source of Variation	Adjusted *p* Value	Summary	Adjusted *p* Value	Summary	Adjusted *p* Value	Summary	Adjusted *p* Value	Summary
Interaction	<0.0001	****	0.0284	*	0.9085	ns	0.0016	**
Hydroxytyrosol	<0.0001	****	<0.0001	****	<0.0001	****	<0.0001	****
Metformin	<0.0001	****	<0.0001	****	<0.0001	****	0.0386	*

2.2. *Hydroxytyrosol and Metformin Synergistically Reduce Aggressiveness Parameters of Prostate Cancer Cells*

To further explore the synergic effect of HT + metformin in PCa cells, we treated LNCaP cells with HT (10 µM) and metformin (2.5 mM), alone or in combination, as well as with the control vehicle for 24 h and performed migration and tumoursphere growth assays. Both individual treatments significantly reduced the migration rate of LNCaP cells (Figure 2A). Interestingly, the combination of HT + metformin potentiated the effect of the individual treatments, reducing the migration rate by more than 50% compared to the vehicle-treated cells (Figure 2A). In the tumoursphere growth assay, we observed a similar behaviour but, in this case, only the combination of HT + metformin significantly reduced this functional parameter in LNCaP cells (Figure 2B). Individually, HT and metformin moderately reduced tumoursphere growth but these reductions did not reach statistical significance (Figure 2B).

Thus, to further explore the cellular response to HT and metformin, we decided to perform apoptosis (early and late apoptosis, performed at 6 h and 48 h after treatment, respectively) and cell cycle (by flow cytometry) assays. In the apoptosis assay, after 6 h of the treatments (early apoptosis), there were no differences relative to the control vehicle (Figure 2C). In contrast, when we evaluated late apoptosis induction (treatments incubated for 48 h), metformin alone and the combination of HT + metformin were able to induce apoptosis, but the combination appeared to show a higher rate of apoptosis stimulation than metformin alone, although this difference was not significant (Figure 2D). Next, we found that the combination of HT and metformin provoked a significant increase in cell arrest at the G1 phase of the cell cycle compared to the control and individual treatments in LNCaP cells, resulting in a significant reduction in the proportion of cells in the S/M phase compared to the control and individual HT treatments, and in the G2 phase compared to the individual HT treatment (Figure 2E). Metformin alone was also able to provoke cell arrest at the G1 phase compared to the vehicle-treated cells. Altogether, our in vitro results confirmed the potential functional anti-tumour effects of the combination of HT and metformin in the LNCaP cell model.

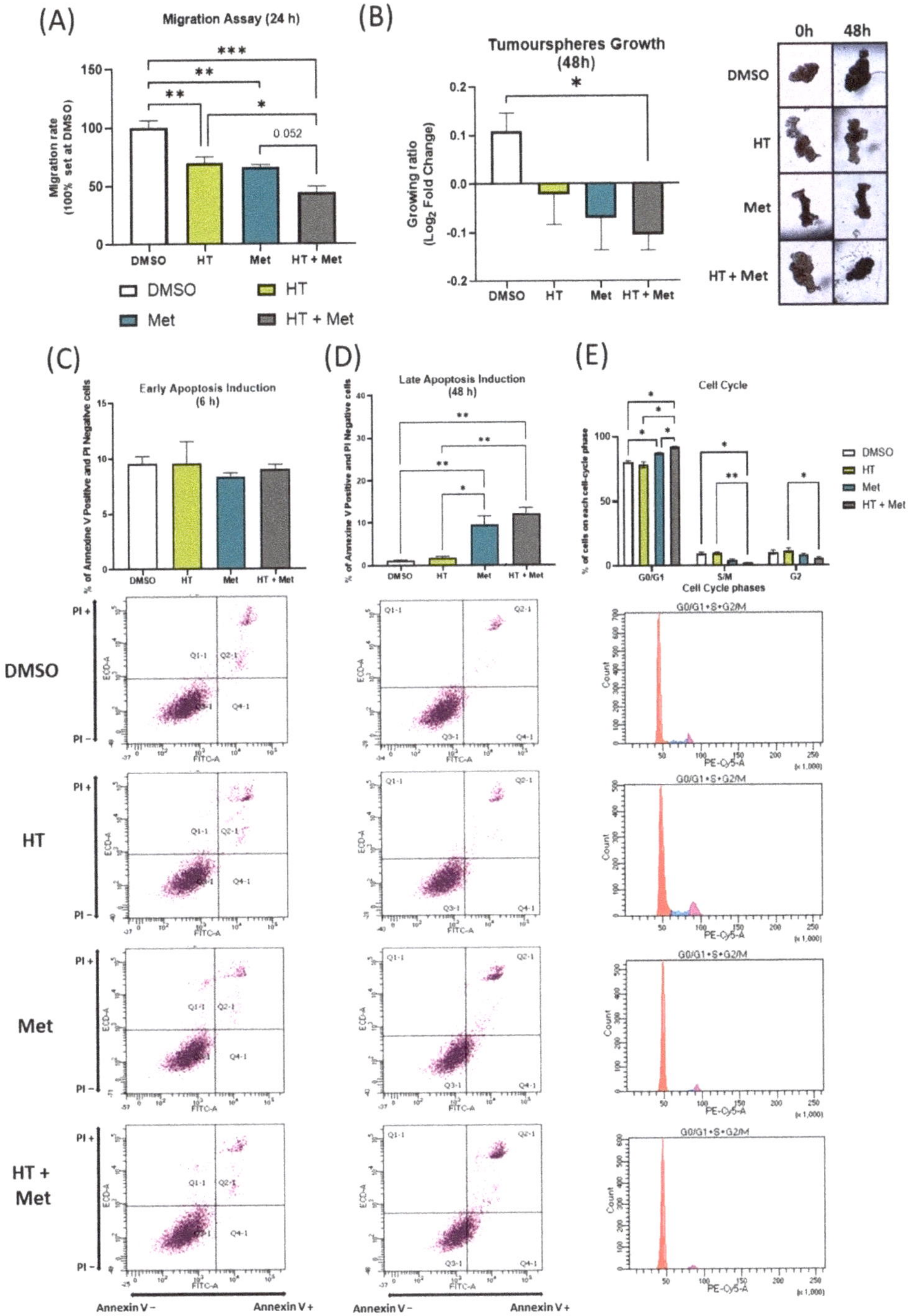

Figure 2. Functional effect of treatment with combination of hydroxytyrosol and metformin. (**A**) Cell migration rate of LNCaP cells treated for 24 h with 10 μM HT, 2.5 mM Met, and their combination.

Data are expressed as percentage of migration of control cells (set at 100%) and represent the mean ± SEM ($n \geq 3$). (**B**) Tumoursphere growth of LNCaP cells after already-formed tumourspheres were treated for 48 h with 10 µM HT, 2.5 mM Met, and their combination. (**C**) Early (at 6 h) and (**D**) late apoptosis (at 48 h) were determined by flow cytometry using Annexin V and PI staining. Data represent cells with positive Annexin V and negative PI staining. (**E**) Percentage of cells in each cell cycle phase, determined using the Watson Pragmatic model. Asterisks indicate significant differences compared to DMSO controls, which were set at 100%. Red represents phase G0/G1, blue represents phase M/S, and purple represents phase G2. Tendency to reach statistical significance is represented by *p*-values and asterisks (* $p < 0.05$, ** $p < 0.01$, and *** $p < 0.001$) indicate statistically significant differences between groups. Abbreviations: HT, hydroxytyrosol; Met, metformin; HT + Met, combination of hydroxytyrosol and metformin.

2.3. Combination of Hydroxytyrosol and Metformin Altered Relevant Molecular Pathways in Prostate Cancer Cells

We used a phosphorylation array to interrogate the potential molecular mechanisms underpinning the functional anti-tumour effect of the combination of HT + metformin in the LNCaP cell model. With this approach, we simultaneously evaluated the phosphorylation of 47 proteins implicated in four different oncogenic pathways (specifically the MAPK, AKT/mTOR, JAK/STAT, and TGF-b pathways; Figure 3A). In general, we found that the combination HT + metformin and the individual treatments might exert an anti-tumour effect through the down-phosphorylation of these signalling pathways (Figure 3B). Specifically, the main effectors of the MAPK signalling pathway, ERK1 and ERK 2, showed a down-phosphorylation in response to the combination of HT + metformin compared to the individual treatments, while other factors (i.e., MKK6 and CREB) showed a similar downregulation in response to the combination and individual treatments (Figure 3C). Additionally, the metformin treatment seemed to be more effective in altering the AKT/mTOR signalling pathway, as we observed stronger inhibitory effects of the metformin treatment alone compared to the combination treatment of HT + metformin (Figure 3B,D). However, the phosphorylation of some upstream (i.e., PDK1) and downstream (i.e., BAD, GSK3b, 4E-BP1, and p70S6K) factors were more downregulated in response to the combination treatment of HT + metformin compared to the individual treatments (Figure 3D).

Furthermore, we found that the JAK/STAT signalling pathway was not drastically altered in response to the HT and/or metformin treatments, and only EGFR was slightly down-phosphorylated in response to the combined HT + metformin treatment compared to the individual treatments (Figure 3E), while some proteins of the STAT family (i.e., Stat2 and Stat3) were slightly over-phosphorylated (Figure 3E). Remarkably, our results revealed that the main inhibitory effects of the combined treatment with HT + metformin seemed to be associated with the TGF-b pathway, since this combination was able to markedly reduce the phosphorylation levels of ATF2, SMAD1, SMAD2, and SMAD5 (key components in the phosphorylation cascade), and also c-Fos and c-Jun (transcription factors directly related to proliferation, cell cycle arrest, and apoptosis) compared to the individual treatments (Figure 3F). Moreover, given the relevance of the AR signalling pathway in PCa pathophysiology, we decided to measure both *AR* mRNA levels (Figure 3G) and *AR* activity (using the signature described by Spatt. et al. [31]; Figure 3H) in response to HT and/or metformin but no differences were found in the treated vs. non-treated (control) cells.

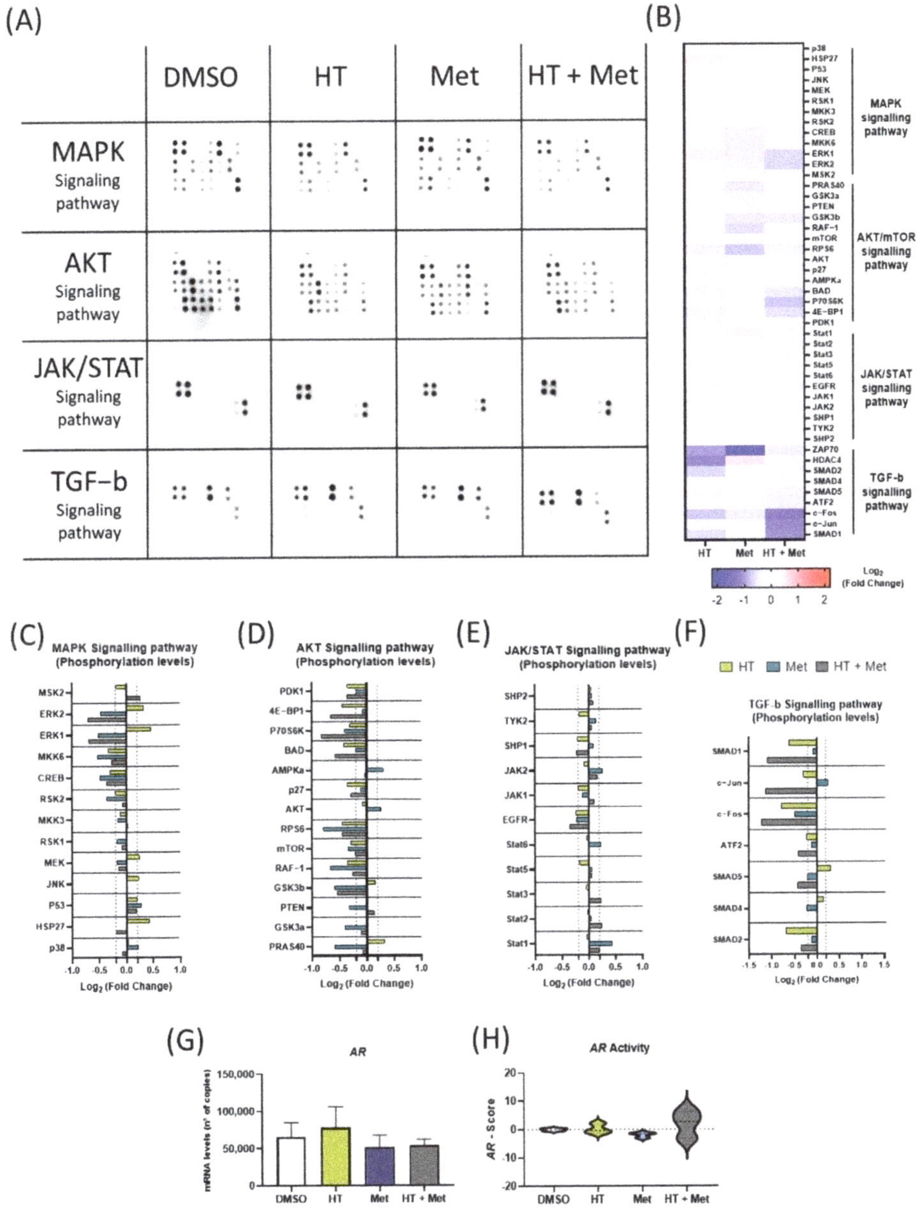

Figure 3. Effect of HT, Met, and their combination on the phosphorylation of key proteins involved cancer-related pathways in LNCaP cells. (**A**) Phosphoprotein array membranes showing the quantified spots reflecting the phosphorylation levels of 47 proteins belonging to four oncogenic pathways (MAPK, AKT, JAK/STAT, and TGF-b) in response to 48 h of treatment with HT, Met, and their combination ($n = 6$; pooled). (**B**) Heatmap showing the mean log2 fold change of the phosphorylation signal corresponding to each depicted protein. ((**C**–**F**)) log2 fold change of phosphorylated protein levels in comparison with the control (DMSO) condition. The dotted line indicates a fold change of ± 0.2. *AR* mRNA levels (**G**) and AR activity (**H**) were measured in treated LNCaP cells. Abbreviations: HT, hydroxytyrosol; Met, metformin; HT + Met, combination of hydroxytyrosol and metformin.

3. Discussion

PCa is a significant global health concern due to its widespread occurrence and high death rate [1]. Unfortunately, the current treatments (surgery, hormone therapy, radiotherapy, and chemotherapy) fail for many patients [3]. This leads to a decline in quality of life for the patients and their families, especially for those diagnosed at later stages, and generates substantial healthcare costs, thereby reinforcing the necessity to find novel therapeutic approaches that can be fast-tracked into the clinic. An emerging therapeutic approach in the oncology field is repositioning drugs that are already approved for other pathologies [4]. In this scenario, recent studies from our group have demonstrated that drugs used to treat metabolic disorders (such as metformin), as well as bioactive compounds within the Mediterranean diet [i.e., those derived from edible vegetables or present in extra virgin olive oil (EVOO), such as hydroxytyrosol (HT; the main phenolic antioxidant in EVOO)], exert anti-tumour actions in PCa [13–16,28,32,33]. These separate findings, together with the potential benefits of combining metformin with natural antioxidants reported by other groups [29], prompted us to evaluate whether the combination of metformin and HT could translate into strong beneficial anti-tumour actions in PCa cells compared to the individual treatments. Therefore, to the best of our knowledge, the present report provides the first comprehensive analysis of a direct side-by-side comparison of the anti-tumour effects that metformin, HT, and the combination of both compounds exert on different human PCa cell models [androgen-sensitive (LNCaP and 22Rv1 cells), androgen-independent (DU 145 cells), and CRPC (PC−3 cells) models], and of their possible underlying mechanisms (i.e., modulation of the activity of multiple key oncogenic cellular processes and signalling pathways that are critical for tumour development, progression, and aggressiveness in human PCa).

First, we performed a dose–response analysis based on a proliferation assay and the SynergicFinder method [30] in the LNCaP, 22Rv1, DU 145, and PC−3 cell models. This approach revealed that LNCaP cells were the cell model exhibiting the highest sensitivity (synergistic response) and a significant interaction effect in response to the combination of HT and metformin. Moreover, and as previously reported, treatment with HT alone mainly affected 22Rv1 cells [28], while metformin alone affected all the PCa cell lines [11,16]. Remarkably, we demonstrated that the anti-proliferative effect of the co-treatment with HT + metformin was significantly more pronounced than the individual treatments with HT or metformin (used at the same concentration as in the combination) in LNCaP cells. In line with these data, we also demonstrated that the combination of HT and metformin enhanced the observed anti-migratory effect as well as the reduction in tumoursphere growth compared to the individual treatments in LNCaP cells. Importantly, it should be emphasized that among all the established human PCa cell lines, LNCaP cells are unique in their ability to model key aspects of PCa progression, including AR signalling and responses to therapeutic agents [34,35]. Therefore, these overall results might be therapeutically relevant and suggest that a well-designed clinical trial is warranted to investigate the effects of the combination of HT + metformin to reduce key clinical aggressiveness features in PCa patients. In line with this, previous studies have also proven that the combination therapy of metformin and different plant-based bioactive compounds or even with other therapies (i.e., chemotherapy, radiotherapy, targeted therapy, and immunotherapy) might exert more significant anti-tumour and/or clinical beneficial effects than monotherapy in different cancer types [36–42]. However, to the best of our knowledge, no other studies have reported similar patterns of responses to the combination of metformin and EVOO bioactive compounds, specifically HT, in PCa or in other endocrine-related cancers. Therefore, these findings emphasize the innovative nature of this approach and highlight the need for further research to evaluate the potential utility of these drugs, particularly their combination, as an effective therapeutic tool for the management of different human tumour pathologies, including PCa.

Next, to further explore the cellular outcomes that underlie the synergistic functional effects observed in response to the combination of HT and metformin in PCa cells, we evaluated two functional and molecular endpoints, apoptosis and cell cycle arrest, that are critical in the pathophysiology of cancer cells. This approach revealed that the combination of HT + metformin increased the late apoptosis of LNCaP cells due to cell cycle arrest. Thus, these findings suggest that the proposed drug combination of HT + metformin holds potential clinical relevance in PCa cells with characteristics similar to those of LNCaP cells, which demonstrated significant sensitivity in our study. However, further studies are necessary to evaluate the applicability of these findings in other PCa subtypes to better understand its broader clinical potential [43]. In support of these findings, previous studies have reported a similar enhancement effect with the combination of metformin and other bioactive antioxidant compounds (i.e., resveratrol) in other cancer cells, including breast cancer [MCF-7 [44] and MDA-MB-23 [45]) and pancreatic carcinoma [46]. Furthermore, it has been also reported that the combination of metformin with curcumin (a diarylheptanoid from the roots of turmeric (Curcuma longa)) is able to increase apoptosis in LNCaP cells, but cell cycle arrest was not affected by this combination [47]. Overall, our data demonstrated that the combination therapy of HT + metformin not only exerted a significantly more pronounced increase in apoptosis through cell cycle arrest, but also significantly reduced proliferation, migration, and tumoursphere formation compared to the individual treatments in LNCaP cells, which should be potentially considered as a novel anti-tumoural avenue to treat PCa patients.

Finally, to interrogate the signalling mechanisms underlying the synergistic anti-tumour actions of the combination of HT and metformin in PCa cells, we analysed the activity of different oncogenic signalling pathways that are important in PCa pathophysiology. Interestingly, we found that the combination of HT + metformin altered the phosphorylation pattern of different critical signalling pathways in PCa cells in a more pronounced way than the individual treatments with HT or metformin, including a further down-phosphorylation of ERK1 and ERK2 (from the MAPK pathway), P70S6K and 4E-BP1 (from the AKT pathway), as well as c-Fos, c-Jun, ATF2, and SMAD1/5 (from the TGF-b pathway). Modulation of the ERK pathway has been associated with PCa progression [48] and it has been widely reported that the inhibition of ERK activity can prevent tumour growth, induce apoptosis and cell cycle arrest, and reduce metastasis in PCa [48–50]; similar effects were observed in other cancer types such as breast [51] and bladder cancer [52]. Likewise, P70S6K and 4E-BP1 are effectors of mTOR and an increase in their activity is associated with the up-regulation of the transcription of genes involved in enhancing cell proliferation and tumourigenicity [53]. The inhibition of P70S6K has been demonstrated to reduce cell proliferation, directly affecting the ability of tumours to grow and maintain themselves [54]. Inhibition of 4E-BP1 induces an anti-tumour response in prostate [55,56] and colorectal cancer [57]. Furthermore, c-Fos and c-Jun modulation is involved in the development of aggressive phenotypes of PCa (from an androgen-dependent to androgen-independent phenotype) [58], wherein a reduction in c-Jun activation has been related to a lower migrative phenotype, which is associated with regulation from SMAD proteins [59]. Additionally, both HT and the combination of HT + metformin reduced the phosphorylation levels of SMAD2, which is a regulator of c-Jun in the canonical TGF-b pathway in PCa cells and is correlated with a poorer prognosis [60]. On the other hand, a general degradation of ATF2 is associated with the suppression of the aggressive features of PCa [61]. Therefore, these data provide original, compelling evidence that the combination of HT + metformin is functionally linked to these well-known, patho-physiologically relevant, oncogenic pathways in PCa (the MASPK, AKT, and TGF-b pathways). Most of the changes observed in the activity of these oncogenic signalling cascades could explain the aforementioned synergic anti-tumour actions of the combination of HT + metformin in terms of increasing apoptosis and inhibiting proliferation, migration, and tumoursphere formation compared to the individual treatments in LNCaP cells. Finally, we did not find any significant changes in the expression or activity of AR (a key oncogenic factor associated with PCa pathophysiology and linked to several of the dysregulated signalling pathways observed in this study [62]) in response to HT and/or metformin. It should be mentioned that previous studies have reported modulation of AR expression in response to metformin; these divergent observations between our study and previous studies might be explained in part by the short exposure time used in our study compared with other studies [63].

In conclusion, this study provides solid evidence indicating that the combination of HT + metformin synergistically reduces the aggressiveness features of androgen-sensitive PCa cells (i.e., decrease in proliferation rate, migration capacity, and/or tumoursphere formation, and activation of apoptosis and cell cycle arrest) compared with the individual treatments with both drugs; these actions are likely mediated by a more pronounced alteration of distinct key oncogenic signalling pathways (i.e., the MAPK, AKT, and TGF-b pathways) that may result in these synergistic effects. Although further work will be required to complete our understanding of this cellular process and to fully elucidate the translational potential behind these interesting and potentially relevant observations, this study suggests the use of these drugs, especially their combination, as a potential new therapeutic tool for the management of PCa that should be tested soon for their use in humans through a well-designed clinical trial.

4. Materials and Methods

4.1. Cell Cultures and Reagents

LNCaP, 22Rv1, DU 145, and PC−3 cell lines were obtained from the ATCC (CRL-1740, CRL-2505, HTB-81, and CRL-1435 respectively), cultured, and maintained under the manufacturer's recommendations, as previously reported [28,32]. Briefly, RPMI 1640 (Corning, Glendale, Arizona, USA; Ref: 15-040-CV) was used and supplemented with 1 % L-glutamine (BIOWEST, Nuaillé, France; Ref: X0550-100), 1% antibiotic–antimycotic solution (Sigma-Aldrich, Madrid, Spain; Ref: A5955-100ML), and 10% FBS (Gibco, Thermo-Fisher, Waltham, MA, USA; Ref: 10270106). The cell lines were validated by analysing short tandem repeats (GenePrint 10 System; Promega, Madison, USA; Ref: B9510) and they were systematically checked for mycoplasma contamination, as previously described [64].

3-Hydroxytyrosol (HT; Ref: 91404) and metformin (Ref: D150969) were obtained from Sigma-Aldrich. Stock solutions were prepared by dissolving the HT powder in dimethyl sulfoxide [DMSO (Applichem, Chicago, USA; Ref: A3672-0100)] and the metformin powder was dissolved in a phospho-buffer solution (PBS). The final DMSO concentration was 0.1% (*v*/*v*). DMSO and PBS were dissolved in culture medium and used as the control vehicle.

4.2. Synergy Assay Design and Analysis

SynergyFinder software [30] was used to evaluate the interaction between HT and metformin in PCa cells. To that end, five concentrations (see Table 1) were selected for each reagent based on the effective concentrations that were described in our previous papers [8,11,16,28]. A proliferation assay was performed to evaluate all the possible combinations. Then, the Loewe additivity model [30,65] was used to calculate the Combinatory Synergy Score for every combination (CSS). Following the SynergyFinder instructions, the CSS was interpreted as follows: (i) antagonistic effect when the CSS is < −10; (ii) additive effect when the CSS is between −10 and 10; and (iii) synergistic effect when the CSS is > 10.

4.3. Cell Proliferation, Migration, and Tumoursphere Growth Assays

Cell proliferation was evaluated in LNCaP, 22Rv1, DU 145, and PC−3 cells using the resazurin assay (Canvax Biotech, Cordoba, Spain; Ref: CA035S), as previously described [11,66]. Briefly, cells were seeded into 96-well plates at a density of 3000 cells/well. Then, resazurin was added to each well to a final concentration of 10%, followed by 3 h of incubation in a humidified incubator at 37 °C, and then the fluorescence (560/590 nm) was measured using a FlexStation III system (Molecular Devices, Sunnyvale, CA, USA). Cell proliferation was determined at 0 and 48 h after treatments with HT and/or metformin.

Cell migration was evaluated using Millicell Cell Culture Inserts (Sigma Aldrich, Ref: PI8P01250) following the manufacturer's instructions. Briefly, 30,000 cells pre-treated with HT and/or metformin for 24 h were resuspended in 300 µL of serum-free RPMI 1640 medium and seeded into the transwell inserts, which were placed in the wells of a 24-well plate with 300 µL complete medium (10% FBS). The cells were incubated at 37 °C for 24 h, and the migrated cells were stained and fixated with crystal violet (0.05% crystal violet and 6% glutaraldehyde). Then, after fixation, the crystal violet was recovered from cells using 10% acetic acid, and the absorbance at 560 nm was determined using a spectrophotometer. A negative control with serum-free medium in the lower chamber was used in each experiment.

Tumoursphere formation was also analysed in response to the HT and/or metformin treatments, as previously described [66,67]. Briefly, 2000 cells/well were seeded into Corning Costar 24-well ultra-low attachment plates (Corning; Ref: CLS3473) with DMEM F-12 (Gibco; Ref: CLS3473) medium supplemented with 20 ng/mL EGF (Sigma-Aldrich; Ref: E9644), 10 ng/mL FGF (Peprotech, London, UK; Ref: 100-18B), and B27 (Thermo-Fisher, Waltham, MA, USA; Ref: 12587010). All supplements were replenished every three days. After 10 days, once tumourspheres were formed, the cells were treated with HT and/or metformin. Pictures were taken at 0 and 48 h after treatment. Images were analysed by using FIJI software v1.54 [68]. The tumourspheres' growth was calculated as a fold change by dividing the area of tumourspheres at 48 h by that at 0 h.

4.4. Flow Cytometry (Apoptosis and Cell Cycle Assays)

Flow cytometry was used to analyse the cell cycle and apoptosis features as previously described [8]. Briefly, LNCaP cells were seeded into 6-well plates (∼2 × 105 cells/well) and treated with HT and/or metformin for 6 h (to evaluate early apoptosis) and 48 h (to evaluate late apoptosis). The cells were trypsinized and then washed with PBS. To evaluate the cell cycle, 200,000 cells were fixed with ice-cold 70% ethanol at 4 °C. The FITC Annexin V Apoptosis Detection Kit I (BD Pharmingen, San Jose, CA, USA; Ref: 556547) and 7-aminoactinomycin D (7-AAD, Invitrogen, Waltham, MA, USA; Ref: A1310) in PBS were used to quantify apoptosis and cell cycle arrest, respectively, according to the manufacturer's instructions. To determine the percentage of apoptotic cells, Annexin V+/Propidium iodide (PI)− cells were designated as early apoptotic cells, whereas Annexin V+/PI + cells were identified as late apoptotic cells. The percentages of cells at the G0/G1, S, and G2/M phases of the cell cycle were calculated by the Watson Pragmatic mode [69]. Cell debris and aggregates were excluded using a sequential gating strategy based on cell size and granularity as previously described [8]. The cells were analysed using a BD LSRFortessa SORP flow cytometer Software v9.0 (BD Biosciences, San Jose, CA, USA; Ref: SCR_018655).

4.5. Phosphorylation Array

Protein extracts of LNCaP cells were collected in lysis buffer from 6-well plates after 2 days of treatment with DMSO and PBS dissolved in the culture medium (control), HT (10 μM), metformin (2.5 mM), or HT-Met (selected combination doses). An exploratory analysis of altered signalling pathways was performed by using the Human Phosphorylation Pathway Profiling Array C55 kit, following the manufacturer's instructions (Raybiotech, Peachtree Corners, GA, USA; Ref: #AAH-PPP-1) and as previously reported [8,28]. Briefly, membranes for the semi-quantitative detection of 55 phosphorylated human proteins belonging to the MAPK, AKT, JAK/STAT, NF-κB, and TGF-β signalling pathways were incubated for 30 min with blocking buffer at 25 °C and then incubated overnight at 4 °C with 1 mL of a 4-fold dilution of the LNCaP cell lysates ($n = 6$; pooled). After washing, the membranes were incubated with a detection antibody cocktail at 25 °C for 2 h and then with a horseradish peroxidase (HRP)-labelled anti-rabbit secondary antibody at 25 °C for an additional 2 h. The signals were collected after adding an ECL reagent using a chemiluminescence detection system (BioRad Universal Hood II, BioRad Laboratories). Densitometric analysis of the array spots was performed using FIJI software [68]. Positive control spots were used as a normalizing factor. The results are expressed as log2 of the fold change of the signal of each protein compared with the signal of the control. The NF-κB results were discarded because of unusable signals.

4.6. RNA Isolation, Retrotranscription, qPCR, and AR-Activity Calculation

RNA was isolated from LNCaP cells treated with HT and/or metformin (48 h) using the TRIzol reagent (Thermo Fisher Scientific), as previously described [66]. The RNA was treated using the RNase-Free DNase Kit (QIAGEN, Shanghai, China) to remove DNA. A Nanodrop One Spectrophotometer (Thermo Fisher Scientific) was used to determine the total RNA concentration and purity. cDNA was synthesized from the total RNA using the cDNA First Strand Synthesis Kit (Thermo Fisher Scientific) and random hexamer primers. Real-time quantitative PCR (qPCR) was performed using a CFX Duet Real-Time PCR System (Bio-Rad, Hercules, CA, USA) with the iTaq Universal SYBR Green Supermix (Bio-Rad). Normalization was performed using a normalization factor calculated by GeNorm 3.3 software [70] using *ACTB* and *GAPDH* expression levels, as previously reported [67]. The AR signalling activity (AR-score) was determined as the sum of the ranked expression levels of eight canonical AR-regulated genes that represent markers of AR activity (*ACSL3, FKBP5, KLK2, KLK3, NKX3-1, PLPP1, RAB3B*, and *STEAP1*) [31], as previously described [66]. The primers used in this study can be found in Appendix A—Table A2.

4.7. Statistical Analysis

Two-way ANOVA was performed to determine the statistical interaction between HT and Met (results shown in Table 2). All possible comparisons between all the combinatorial doses were performed against the negative control using Dunnett's multiple comparison test (results shown in Supplemental Table 1). Statistical differences between two independent conditions were determined using unpaired parametric t-tests or nonparametric Mann–Whitney U tests based on the normality, which was assessed using the Kolmogorov–Smirnov test. The assays were performed in at least 3 independent experiments ($n \geq 3$) and with at least 2 technical replicates. $p < 0.05$ was considered statistically significant. A trend for significance was indicated when the p-values ranged between >0.05 and <0.1. In the Human Phosphorylation Pathway Profiling Array, a fold change (log2) of 0.2 was used as the significance threshold. All the analyses were performed using GraphPad 9.4.1 software.

Author Contributions: F.P.-P.: Software, Validation, Formal Analysis, Investigation, Writing, Visualization. A.J.M.-H.: Investigation, Writing—Review and Editing. M.E.G.-G.: Investigation, Writing—Review and Editing. I.G.-D.: Validation, Writing—Review and Editing. A.P.-E.: Validation, Writing—Review and Editing. M.D.G.: Writing—Review and Editing. A.S.-C.: Writing—Review and Editing. R.M.L.: Conceptualization, Writing—Review and Editing, Supervision, Project Administration, Funding Acquisition. A.J.L.-G.: Conceptualization, Writing—Review and Editing, Supervision, Project Administration, Funding Acquisition. All authors have read and agreed to the published version of the manuscript.

Funding: This work was funded by MICINN (PID2022-1381850B-I00; FPU18/02485; PRE2020-094225; FPU20/03954), Universidad de Córdoba (FPU23/02246, PRE2022-000741), Junta de Andalucía (BIO-0139, PI-0152-2019), and CIBERobn. CIBER is an initiative of the Instituto de Salud Carlos III, Ministerio de Sanidad, Servicios Sociales e Igualdad, Spain.

Institutional Review Board Statement: The study was conducted in accordance with the Declaration of Helsinki, and approved by the Reina Sofia University Hospital Ethics Committee (5570, approved on 29 May 2023).

Informed Consent Statement: Not applicable.

Data Availability Statement: Not applicable.

Conflicts of Interest: The authors declare no conflicts of interest.

Abbreviations

The following abbreviations are used in this manuscript:

PCa	Prostate cancer
EVOO	Extra virgin olive oil
HT	Hydroxytyrosol
Met	Metformin
HT + Met	Combination of hydroxytyrosol and metformin
CRPC	Castration-Resistant Prostate Cancer

Appendix A

Table A1. Statistics for drug combination effects in cell proliferation assay. Statistical differences (*p*-values and asterisks) are represented for all dose combinations of each treatment (ns, $p > 0.05$; *, $p < 0.05$; **, $p < 0.01$; ***, $p < 0.001$; and ****, $p < 0.0001$).

Drugs Dose Combination		LNCaP		22Rv1		DU 145		PC-3	
Hydroxytyrosol (µM)	**Metformin (mM)**	**Adjusted *p* Value**	**Summary**	**Adjusted *p* Value**	**Summary**	**Adjusted *p* Value**	**Summary**	**Adjusted *p* Value**	**Summary**
0	0.5	0.9996	ns	0.8065	ns	>0.9999	ns	0.0042	**
0	1	0.9877	ns	0.9400	ns	>0.9999	ns	0.9996	ns
0	2.5	0.9996	ns	0.6923	ns	0.8925	ns	0.6425	ns
0	5	<0.0001	****	0.9991	ns	0.4907	ns	0.9863	ns
3	0	0.7956	ns	0.8360	ns	>0.9999	ns	0.0088	**
3	0.5	0.8337	ns	0.1204	ns	0.9993	ns	0.0381	*
3	1	0.3196	ns	0.4763	ns	0.9992	ns	0.0025	**
3	2.5	0.0167	*	0.0839	ns	0.9521	ns	0.0215	*
3	5	<0.0001	****	0.9996	ns	0.9990	ns	<0.0001	****

Table A1. *Cont.*

Drugs Dose Combination		LNCaP		22Rv1		DU 145		PC-3	
Hydroxytyrosol (μM)	Metformin (mM)	Adjusted *p* Value	Summary	Adjusted *p* Value	Summary	Adjusted *p* Value	Summary	Adjusted *p* Value	Summary
10	0	0.9274	ns	0.9998	ns	0.9871	ns	0.9228	ns
10	0.5	0.9998	ns	0.9952	ns	0.9997	ns	0.0019	**
10	1	0.9941	ns	0.9944	ns	0.9990	ns	0.0001	***
10	2.5	0.0051	**	0.7021	ns	0.9996	ns	0.7973	ns
10	5	<0.0001	****	0.9342	ns	0.5193	ns	0.1046	ns
30	0	0.0192	*	0.0054	**	0.9996	ns	0.0051	**
30	0.5	0.1979	ns	0.0023	**	>0.9999	ns	<0.0001	****
30	1	0.1942	ns	0.0016	**	>0.9999	ns	<0.0001	****
30	2.5	0.1976	ns	0.9996	ns	0.9614	ns	<0.0001	****
30	5	<0.0001	****	0.0379	*	0.7128	ns	<0.0001	****
100	0	<0.0001	****	<0.0001	****	0.0126	*	0.0005	***
100	0.5	<0.0001	****	<0.0001	****	0.5459	ns	<0.0001	****
100	1	<0.0001	****	<0.0001	****	0.1556	ns	<0.0001	****
100	2.5	<0.0001	****	<0.0001	****	0.0021	**	<0.0001	****
100	5	<0.0001	****	<0.0001	****	<0.0001	****	<0.0001	****

Table A2. Specific primers for the human genes used in this study. The official names of the genes, their accession numbers, the primer sequences, and the product sizes of the amplification products are included.

Gene	Ensembl Ref	Sense Primer Sequence	Antisense Primer Sequence	Product Size (bp)
ACTB	ENSG00000075624	ACTCTTCCAGCCTTCCTTCC	CAGTGATCTCCTTCTGCATC	176
GAPDH	ENSG00000111640	GCCTCAAGATCATCAGCAATG	CTTCCACGATACCAAAGTTGT	90
AR	ENSG00000169083	GCAGGAAGCAGTATCCGAAG	GTTGTCAGAAATGGTCGAAGTG	112
ACSL3	ENSG00000123983	TTCCAGAACTAGGAGAGGAAGATG	CATCCGTGAGAAAGACAGACAA	97
FKBP5	ENSG00000096060	CGGAGAACCAAACGGAAA	TCAAACATCCTTCCACCACA	98
KLK2	ENSG00000167751	TGATTCTGGGGGTCCACTT	GTACACAGCAGGCTTTTCAGG	97
KLK3	ENSG00000142515	AGAGGAGTTCTTGACCCCAAA	CCTTCTGAGGGTGAACTTGC	89
NKX3-1	ENSG00000167034	TCAGGTGATCGAGTTGGAGAG	TCCGTGAGCTTGAGGTTCTT	93
PLPP1	ENSG00000067113	CCTCACTTCTTGGATGTTTGTG	GACAACCTGCCTTCCTTAACTCT	116
RAB3B	ENSG00000169213	TCCTCTTCCGCTATGCTGAT	AGTTTCACCCGCTTCTCGT	106
STEAP1	ENSG00000164647	TGGCAATACTGGCTCTGTTG	GGGAAACAATTCCTAGCTTGC	105

References

1. Bray, F.; Laversanne, M.; Sung, H.; Ferlay, J.; Siegel, R.L.; Soerjomataram, I.; Jemal, A. Global Cancer Statistics 2022: GLOBOCAN Estimates of Incidence and Mortality Worldwide for 36 Cancers in 185 Countries. *CA Cancer J. Clin.* **2024**, *74*, 229–263. [CrossRef] [PubMed]
2. Mukherji, D.; Omlin, A.; Pezaro, C.; Shamseddine, A.; de Bono, J. Metastatic Castration-Resistant Prostate Cancer (CRPC): Preclinical and Clinical Evidence for the Sequential Use of Novel Therapeutics. *Cancer Metastasis Rev.* **2014**, *33*, 555–566. [CrossRef] [PubMed]
3. Tyson, M.D.; Penson, D.F.; Resnick, M.J. The Comparative Oncologic Effectiveness of Available Management Strategies for Clinically Localized Prostate Cancer. *Urol. Oncol.* **2017**, *35*, 51–58. [CrossRef]

4. Pushpakom, S.; Iorio, F.; Eyers, P.A.; Escott, K.J.; Hopper, S.; Wells, A.; Doig, A.; Guilliams, T.; Latimer, J.; McNamee, C.; et al. Drug Repurposing: Progress, Challenges and Recommendations. *Nat. Rev. Drug Discov.* **2019**, *18*, 41–58. [CrossRef]
5. Polesel, J.; Gini, A.; Dal Maso, L.; Stocco, C.; Birri, S.; Taborelli, M.; Serraino, D.; Zucchetto, A. The Impact of Diabetes and Other Metabolic Disorders on Prostate Cancer Prognosis. *J. Diabetes Complicat.* **2016**, *30*, 591–596. [CrossRef] [PubMed]
6. Maj-Hes, A.B.; Mathieu, R.; Özsoy, M.; Soria, F.; Moschini, M.; Abufaraj, M.; Briganti, A.; Roupret, M.; Karakiewicz, P.I.; Klatte, T.; et al. Obesity Is Associated with Biochemical Recurrence after Radical Prostatectomy: A Multi-Institutional Extended Validation Study. *Urol. Oncol. Semin. Orig. Investig.* **2017**, *35*, e1–e460. [CrossRef]
7. Jamnagerwalla, J.; Howard, L.E.; Allott, E.H.; Vidal, A.C.; Moreira, D.M.; Castro-Santamaria, R.; Andriole, G.L.; Freeman, M.R.; Freedland, S.J. Serum Cholesterol and Risk of High-Grade Prostate Cancer: Results from the REDUCE Study. *Prostate Cancer Prostatic Dis.* **2018**, *21*, 252–259. [CrossRef] [PubMed]
8. Fuentes-Fayos, A.C.; G-García, M.E.; Pérez-Gómez, J.M.; Montero-Hidalgo, A.J.; Martín-Colom, J.; Doval-Rosa, C.; Blanco-Acevedo, C.; Torres, E.; Toledano-Delgado, Á.; Sánchez-Sánchez, R.; et al. Metformin and Simvastatin Exert Additive Antitumour Effects in Glioblastoma via Senescence-State: Clinical and Translational Evidence. *EBioMedicine* **2023**, *90*, 104484. [CrossRef]
9. Vázquez-Borrego, M.C.; Fuentes-Fayos, A.C.; Herrera-Martínez, A.D.; Venegas-Moreno, E.; L-López, F.; Fanciulli, A.; Moreno-Moreno, P.; Alhambra-Expósito, M.R.; Barrera-Martín, A.; Dios, E.; et al. Statins Directly Regulate Pituitary Cell Function and Exert Antitumor Effects in Pituitary Tumors. *Neuroendocrinology* **2020**, *110*, 1028–1041. [CrossRef] [PubMed]
10. Vázquez-Borrego, M.C.; Fuentes-Fayos, A.C.; Herrera-Martínez, A.D.; L-López, F.; Ibáñez-Costa, A.; Moreno-Moreno, P.; Alhambra-Expósito, M.R.; Barrera-Martín, A.; Blanco-Acevedo, C.; Dios, E.; et al. Biguanides Exert Antitumoral Actions in Pituitary Tumor Cells Through AMPK-Dependent and -Independent Mechanisms. *J. Clin. Endocrinol. Metab.* **2019**, *104*, 3501–3513. [CrossRef] [PubMed]
11. Jiménez-Vacas, J.M.; Herrero-Aguayo, V.; Montero-Hidalgo, A.J.; Sáez-Martínez, P.; Gómez-Gómez, E.; León-González, A.J.; Fuentes-Fayos, A.C.; Yubero-Serrano, E.M.; Requena-Tapia, M.J.; López, M.; et al. Clinical, Cellular, and Molecular Evidence of the Additive Antitumor Effects of Biguanides and Statins in Prostate Cancer. *J. Clin. Endocrinol. Metab.* **2021**, *106*, E696–E710. [CrossRef] [PubMed]
12. Herrera-Martínez, A.D.; Pedraza-Arevalo, S.; L-López, F.; Gahete, M.D.; Gálvez-Moreno, M.A.; Castaño, J.P.; Luque, R.M. Type 2 Diabetes in Neuroendocrine Tumors: Are Biguanides and Statins Part of the Solution? *J. Clin. Endocrinol. Metab.* **2019**, *104*, 57–73. [CrossRef]
13. León-González, A.J.; Jiménez-Vacas, J.M.; Fuentes-Fayos, A.C.; Sarmento-Cabral, A.; Herrera-Martínez, A.D.; Gahete, M.D.; Luque, R.M. Role of Metformin and Other Metabolic Drugs in the Prevention and Therapy of Endocrine-Related Cancers. *Curr. Opin. Pharmacol.* **2021**, *60*, 17–26. [CrossRef] [PubMed]
14. Spratt, D.E.; Zhang, C.; Zumsteg, Z.S.; Pei, X.; Zhang, Z.; Zelefsky, M.J. Metformin and Prostate Cancer: Reduced Development of Castration-Resistant Disease and Prostate Cancer Mortality. *Eur. Urol.* **2013**, *63*, 709–716. [CrossRef] [PubMed]
15. Hua, Y.; Zheng, Y.; Yao, Y.; Jia, R.; Ge, S.; Zhuang, A. Metformin and Cancer Hallmarks: Shedding New Lights on Therapeutic Repurposing. *J. Transl. Med.* **2023**, *21*, 403. [CrossRef] [PubMed]
16. Sarmento-Cabral, A.; L-López, F.; Gahete, M.D.; Castaño, J.P.; Luque, R.M. Metformin Reduces Prostate Tumor Growth, in a Diet-Dependent Manner, by Modulating Multiple Signaling Pathways. *Mol. Cancer Res.* **2017**, *15*, 862–874. [CrossRef]
17. Aboelnaga, E.M.; Aboelnaga, M.M.; Elkalla, H.M. Metformin Addition to Androgen Deprivation Therapy Effect on Cancer Prostate Patients with Type 2 Diabetes. *Diabetes Metab. Syndr.* **2021**, *15*, 102251. [CrossRef]
18. Mahalingam, D.; Hanni, S.; Serritella, A.V.; Fountzilas, C.; Michalek, J.; Hernandez, B.; Sarantopoulos, J.; Datta, P.; Romero, O.; Pillai, S.M.A.; et al. Utilizing Metformin to Prevent Metabolic Syndrome Due to Androgen Deprivation Therapy (ADT): A Randomized Phase II Study of Metformin in Non-Diabetic Men Initiating ADT for Advanced Prostate Cancer. *Oncotarget* **2023**, *14*, 622–636. [CrossRef] [PubMed]
19. GBD 2015 Obesity Collaborators. Health Effects of Overweight and Obesity in 195 Countries over 25 Years. *N. Engl. J. Med.* **2017**, *377*, 13–27. [CrossRef]
20. López-Guarnido, O.; Urquiza-Salvat, N.; Saiz, M.; Lozano-Paniagua, D.; Rodrigo, L.; Pascual-Geler, M.; Lorente, J.A.; Alvarez-Cubero, M.J.; Rivas, A. Bioactive Compounds of the Mediterranean Diet and Prostate Cancer. *Aging Male* **2018**, *21*, 251–260. [CrossRef] [PubMed]
21. Celano, M.; Maggisano, V.; Lepore, S.M.; Russo, D.; Bulotta, S. Secoiridoids of Olive and Derivatives as Potential Coadjuvant Drugs in Cancer: A Critical Analysis of Experimental Studies. *Pharmacol. Res.* **2019**, *142*, 77–86. [CrossRef]
22. Cruz-Lozano, M.; González-González, A.; Marchal, J.A.; Muñoz-Muela, E.; Molina, M.P.; Cara, F.E.; Brown, A.M.; García-Rivas, G.; Hernández-Brenes, C.; Lorente, J.A.; et al. Hydroxytyrosol Inhibits Cancer Stem Cells and the Metastatic Capacity of Triple-Negative Breast Cancer Cell Lines by the Simultaneous Targeting of Epithelial-to-Mesenchymal Transition, Wnt/β-Catenin and TGFβ Signaling Pathways. *Eur. J. Nutr.* **2019**, *58*, 3207–3219. [CrossRef] [PubMed]

23. El-azem, N.; Pulido-Moran, M.; Ramirez-Tortosa, C.L.; Quiles, J.L.; Cara, F.E.; Sanchez-Rovira, P.; Granados-Principal, S.; Ramirez-Tortosa, M. Modulation by Hydroxytyrosol of Oxidative Stress and Antitumor Activities of Paclitaxel in Breast Cancer. *Eur. J. Nutr.* **2019**, *58*, 1203–1211. [CrossRef] [PubMed]
24. Costantini, F.; Di Sano, C.; Barbieri, G. The Hydroxytyrosol Induces the Death for Apoptosis of Human Melanoma Cells. *Int. J. Mol. Sci.* **2020**, *21*, 8074. [CrossRef] [PubMed]
25. Castejón, M.L.; Montoya, T.; Alarcón-de-la-lastra, C.; Sánchez-hidalgo, M. Potential Protective Role Exerted by Secoiridoids from *Olea europaea* L. in Cancer, Cardiovascular, Neurodegenerative, Aging-Related, and Immunoinflammatory Diseases. *Antioxidants* **2020**, *9*, 149. [CrossRef]
26. Vilaplana-Pérez, C.; Auñón, D.; García-Flores, L.A.; Gil-Izquierdo, A. Hydroxytyrosol and Potential Uses in Cardiovascular Diseases, Cancer, and AIDS. *Front. Nutr.* **2014**, *1*, 18. [CrossRef]
27. Sani, N.S.; Onsori, H.; Akrami, S.; Rahmati, M. A Comparison of the Anti-Cancer Effects of Free and PLGA-PAA Encapsulated Hydroxytyrosol on the HT-29 Colorectal Cancer Cell Line. *Anticancer Agents Med. Chem.* **2022**, *22*, 390–394. [CrossRef]
28. León-González, A.J.; Sáez-Martínez, P.; Jiménez-Vacas, J.M.; Herrero-Aguayo, V.; Montero-Hidalgo, A.J.; Gómez-Gómez, E.; Madrona, A.; Castaño, J.P.; Espartero, J.L.; Gahete, M.D.; et al. Comparative Cytotoxic Activity of Hydroxytyrosol and Its Semisynthetic Lipophilic Derivatives in Prostate Cancer Cells. *Antioxidants* **2021**, *10*, 1348. [CrossRef]
29. Ibrahim, R.S.; Ibrahim, S.S.; El-Naas, A.; Koklesová, L.; Kubatka, P.; Büsselberg, D. Could Metformin and Resveratrol Support Glioblastoma Treatment? A Mechanistic View at the Cellular Level. *Cancers* **2023**, *15*, 3368. [CrossRef]
30. Ianevski, A.; Giri, A.K.; Aittokallio, T. SynergyFinder 2.0: Visual Analytics of Multi-Drug Combination Synergies. *Nucleic Acids Res.* **2020**, *48*, W488–W493. [CrossRef] [PubMed]
31. Spratt, D.E.; Alshalalfa, M.; Fishbane, N.; Weiner, A.B.; Mehra, R.; Mahal, B.A.; Lehrer, J.; Liu, Y.; Zhao, S.G.; Speers, C.; et al. Transcriptomic Heterogeneity of Androgen Receptor Activity Defines a de Novo Low AR-Active Subclass in Treatment Naïve Primary Prostate Cancer. *Clin. Cancer Res.* **2019**, *25*, 6721–6730. [CrossRef] [PubMed]
32. Libero, M.L.; Montero-Hidalgo, A.J.; Recinella, L.; Luque, R.M.; Generali, D.; Acquaviva, A.; Orlando, G.; Ferrante, C.; Menghini, L.; Di Simone, S.C.; et al. The Protective Effects of an Aged Black Garlic Water Extract on the Prostate. *Nutrients* **2024**, *16*, 3025. [CrossRef] [PubMed]
33. Libero, M.L.; Lucarini, E.; Recinella, L.; Ciampi, C.; Veschi, S.; Piro, A.; Chiavaroli, A.; Acquaviva, A.; Nilofar, N.; Orlando, G.; et al. Anti-inflammatory and Anti-hyperalgesic Effects Induced by an Aqueous Aged Black Garlic Extract in Rodent Models of Ulcerative Colitis and Colitis-associated Visceral Pain. *Phytother. Res.* **2024**, *38*, 4177–4188. [CrossRef] [PubMed]
34. Abate-Shen, C.; Nunes de Almeida, F. Establishment of the LNCaP Cell Line–The Dawn of an Era for Prostate Cancer Research. *Cancer Res.* **2022**, *82*, 1689–1691. [CrossRef]
35. van Bokhoven, A.; Varella-Garcia, M.; Korch, C.; Johannes, W.U.; Smith, E.E.; Miller, H.L.; Nordeen, S.K.; Miller, G.J.; Lucia, M.S. Molecular Characterization of Human Prostate Carcinoma Cell Lines. *Prostate* **2003**, *57*, 205–225. [CrossRef]
36. Wang, Y.; Li, J.; Xia, L. Plant-Derived Natural Products and Combination Therapy in Liver Cancer. *Front. Oncol.* **2023**, *13*, 1116532. [CrossRef] [PubMed]
37. Martínez, N.; Herrera, M.; Frías, L.; Provencio, M.; Pérez-Carrión, R.; Díaz, V.; Morse, M.; Crespo, M.C. A Combination of Hydroxytyrosol, Omega-3 Fatty Acids and Curcumin Improves Pain and Inflammation among Early Stage Breast Cancer Patients Receiving Adjuvant Hormonal Therapy: Results of a Pilot Study. *Clin. Transl. Oncol.* **2019**, *21*, 489–498. [CrossRef] [PubMed]
38. Sharma, S.; Cwiklinski, K.; Mahajan, S.D.; Schwartz, S.A.; Aalinkeel, R. Combination Modality Using Quercetin to Enhance the Efficacy of Docetaxel in Prostate Cancer Cells. *Cancers* **2023**, *15*, 902. [CrossRef] [PubMed]
39. Haghshenas, M.; Firouzabadi, N.; Akbarizadeh, A.R.; Rashedinia, M. Combination of Metformin and Gallic Acid Induces Autophagy and Apoptosis in Human Breast Cancer Cells. *Res. Pharm. Sci.* **2023**, *18*, 663–675. [CrossRef] [PubMed]
40. Das, D.; Kanchan, M.; Mitra, A.; Zaky, M.Y.; Pathak, S.; Banerjee, A. A Review on the Efficacy of Plant-Derived Bio-Active Compounds Curcumin and Aged Garlic Extract in Modulating Cancer and Age-Related Diseases. *Curr. Rev. Clin. Exp. Pharmacol.* **2024**, *19*, 146–162. [CrossRef]
41. Adewale, O.O.; Wińska, P.; Piasek, A.; Cieśla, J. The Potential of Plant Polysaccharides and Chemotherapeutic Drug Combinations in the Suppression of Breast Cancer. *Int. J. Mol. Sci.* **2024**, *25*, 12202. [CrossRef] [PubMed]
42. Zhu, L.; Yang, K.; Ren, Z.; Yin, D.; Zhou, Y. Metformin as Anticancer Agent and Adjuvant in Cancer Combination Therapy: Current Progress and Future Prospect. *Transl. Oncol.* **2024**, *44*, 101945. [CrossRef]
43. Westaby, D.; Jimenez-Vacas, J.M.; Padilha, A.; Varkaris, A.; Balk, S.P.; de Bono, J.S.; Sharp, A. Targeting the Intrinsic Apoptosis Pathway: A Window of Opportunity for Prostate Cancer. *Cancers* **2021**, *14*, 51. [CrossRef] [PubMed]
44. Fatehi, R.; Rashedinia, M.; Akbarizadeh, A.R.; Zamani, M.; Firouzabadi, N. Metformin Enhances Anti-Cancer Properties of Resveratrol in MCF-7 Breast Cancer Cells via Induction of Apoptosis, Autophagy and Alteration in Cell Cycle Distribution. *Biochem. Biophys. Res. Commun.* **2023**, *644*, 130–139. [CrossRef]

45. Cheng, T.; Wang, C.; Lu, Q.; Cao, Y.; Yu, W.; Li, W.; Liu, B.; Gao, X.; Lü, J.; Pan, X. Metformin Inhibits the Tumor-Promoting Effect of Low-Dose Resveratrol, and Enhances the Anti-Tumor Activity of High-Dose Resveratrol by Increasing Its Reducibility in Triple Negative Breast Cancer. *Free Radic. Biol. Med.* **2022**, *180*, 108–120. [CrossRef] [PubMed]
46. Zhu, M.; Zhang, Q.; Wang, X.; Kang, L.; Yang, Y.; Liu, Y.; Yang, L.; Li, J.; Yang, L.; Liu, J.; et al. Metformin Potentiates Anti-Tumor Effect of Resveratrol on Pancreatic Cancer by down-Regulation of VEGF-B Signaling Pathway. *Oncotarget* **2016**, *7*, 84190–84200. [CrossRef] [PubMed]
47. Eslami, S.S.; Jafari, D.; Montazeri, H.; Sadeghizadeh, M.; Tarighi, P. Combination of Curcumin and Metformin Inhibits Cell Growth and Induces Apoptosis without Affecting the Cell Cycle in LNCaP Prostate Cancer Cell Line. *Nutr. Cancer* **2021**, *73*, 1026–1039. [CrossRef] [PubMed]
48. El-Haibi, C.P.; Singh, R.; Sharma, P.K.; Singh, S.; Lillard, J.W. CXCL13 Mediates Prostate Cancer Cell Proliferation through JNK Signalling and Invasion through ERK Activation. *Cell Prolif.* **2011**, *44*, 311–319. [CrossRef] [PubMed]
49. Xue, J.; Zhang, Z.; Hu, H. Prostate Cancer Growth Inhibition by 1-(3,5-Dimethylphenyl)–6-Methyl-1H-Pyrazolo [4,3-c]Pyridin-4(5H)-One via Down-Regulation of Phosphorylation PI3K/AKT and STA3/JAK2. *Dokl. Biochem. Biophys.* **2020**, *495*, 347–353. [CrossRef] [PubMed]
50. McCubrey, J.A.; Steelman, L.S.; Chappell, W.H.; Abrams, S.L.; Wong, E.W.T.; Chang, F.; Lehmann, B.; Terrian, D.M.; Milella, M.; Tafuri, A.; et al. Roles of the Raf/MEK/ERK Pathway in Cell Growth, Malignant Transformation and Drug Resistance. *Biochim. Biophys. Acta (BBA)-Mol. Cell Res.* **2007**, *1773*, 1263–1284. [CrossRef] [PubMed]
51. Zhang, Q.; Lu, S.; Li, T.; Yu, L.; Zhang, Y.; Zeng, H.; Qian, X.; Bi, J.; Lin, Y. ACE2 Inhibits Breast Cancer Angiogenesis via Suppressing the VEGFa/VEGFR2/ERK Pathway. *J. Exp. Clin. Cancer Res.* **2019**, *38*, 173. [CrossRef] [PubMed]
52. Zou, F.; Rao, T.; Chen, W.; Song, T.; Li, T.; Hu, W.; Li, L.; Yu, W.; Cheng, F. DUSP2 Affects Bladder Cancer Prognosis by Down-Regulating MEK/ERK and P38 MAPK Signaling Pathways through PTPN7. *Cell. Signal.* **2023**, *112*, 110893. [CrossRef] [PubMed]
53. Laplante, M.; Sabatini, D.M. MTOR Signaling in Growth Control and Disease. *Cell* **2012**, *149*, 274–293. [CrossRef] [PubMed]
54. Roux, P.P.; Blenis, J. ERK and P38 MAPK-Activated Protein Kinases: A Family of Protein Kinases with Diverse Biological Functions. *Microbiol. Mol. Biol. Rev.* **2004**, *68*, 320–344. [CrossRef] [PubMed]
55. Wang, K.; Wang, K.; Ma, Q. The Expression and Significance of P4E-BP1/4E-BP1 in Prostate Cancer. *J. Clin. Lab. Anal.* **2022**, *36*, e24332. [CrossRef]
56. Ding, M.; Van der Kwast, T.H.; Vellanki, R.N.; Foltz, W.D.; McKee, T.D.; Sonenberg, N.; Pandolfi, P.P.; Koritzinsky, M.; Wouters, B.G. The MTOR Targets 4E-BP1/2 Restrain Tumor Growth and Promote Hypoxia Tolerance in PTEN-Driven Prostate Cancer. *Mol. Cancer Res.* **2018**, *16*, 682–695. [CrossRef]
57. Tenesa, A.; Theodoratou, E.; Din, F.V.N.; Farrington, S.M.; Cetnarskyj, R.; Barnetson, R.A.; Porteous, M.E.; Campbell, H.; Dunlop, M.G. Ten Common Genetic Variants Associated with Colorectal Cancer Risk Are Not Associated with Survival after Diagnosis. *Clin. Cancer Res.* **2010**, *16*, 3754–3759. [CrossRef]
58. Edwards, J.; Krishna, N.S.; Mukherjee, R.; Bartlett, J.M.S. The Role of C-Jun and c-Fos Expression in Androgen-Independent Prostate Cancer. *J. Pathol.* **2004**, *204*, 153–158. [CrossRef]
59. Thakur, N.; Hamidi, A.; Song, J.; Itoh, S.; Bergh, A.; Heldin, C.H.; Landström, M. Smad7 Enhances TGF-β-Induced Transcription of c-Jun and HDAC6 Promoting Invasion of Prostate Cancer Cells. *iScience* **2020**, *23*, 101470. [CrossRef] [PubMed]
60. Thakur, N.; Gudey, S.K.; Marcusson, A.; Fu, J.Y.; Bergh, A.; Heldin, C.-H.; Landström, M. TGFβ-Induced Invasion of Prostate Cancer Cells Is Promoted by c-Jun-Dependent Transcriptional Activation of Snail1. *Cell Cycle* **2014**, *13*, 2400–2414. [CrossRef]
61. Ma, J.; Chang, K.; Peng, J.; Shi, Q.; Gan, H.; Gao, K.; Feng, K.; Xu, F.; Zhang, H.; Dai, B.; et al. SPOP Promotes ATF2 Ubiquitination and Degradation to Suppress Prostate Cancer Progression. *J. Exp. Clin. Cancer Res.* **2018**, *37*, 145. [CrossRef] [PubMed]
62. Pencik, J.; Philippe, C.; Schlederer, M.; Atas, E.; Pecoraro, M.; Grund-Gröschke, S.; Li, W.; Tracz, A.; Heidegger, I.; Lagger, S.; et al. STAT3/LKB1 Controls Metastatic Prostate Cancer by Regulating MTORC1/CREB Pathway. *Mol. Cancer* **2023**, *22*, 133. [CrossRef]
63. Wang, Y.; Liu, G.; Tong, D.; Parmar, H.; Hasenmayer, D.; Yuan, W.; Zhang, D.; Jiang, J. Metformin Represses Androgen-dependent and Androgen-independent Prostate Cancers by Targeting Androgen Receptor. *Prostate* **2015**, *75*, 1187–1196. [CrossRef] [PubMed]
64. Herrero-Aguayo, V.; Jiménez-Vacas, J.M.; Sáez-Martínez, P.; Gómez-Gómez, E.; López-Cánovas, J.L.; Garrido-Sánchez, L.; Herrera-Martínez, A.D.; García-Bermejo, L.; MacÍas-González, M.; López-Miranda, J.; et al. Influence of Obesity in the MiRNome: MiR-4454, a Key Regulator of Insulin Response Via Splicing Modulation in Prostate. *J. Clin. Endocrinol. Metab.* **2021**, *106*, E469–E484. [CrossRef]
65. Goldoni, M.; Johansson, C. A Mathematical Approach to Study Combined Effects of Toxicants in Vitro: Evaluation of the Bliss Independence Criterion and the Loewe Additivity Model. *Toxicol. Vitr.* **2007**, *21*, 759–769. [CrossRef]

66. Montero-Hidalgo, A.J.; Jiménez-Vacas, J.M.; Gómez-Gómez, E.; Porcel-Pastrana, F.; Sáez-Martínez, P.; Pérez-Gómez, J.M.; Fuentes-Fayos, A.C.; Blázquez-Encinas, R.; Sánchez-Sánchez, R.; González-Serrano, T.; et al. SRSF6 Modulates Histone-Chaperone HIRA Splicing to Orchestrate AR and E2F Activity in Prostate Cancer. *Sci. Adv.* **2024**, *10*, eado8231. [CrossRef] [PubMed]
67. Jiménez-Vacas, J.M.; Montero-Hidalgo, A.J.; Gómez-Gómez, E.; Sáez-Martínez, P.; Fuentes-Fayos, A.C.; Closa, A.; González-Serrano, T.; Martínez-López, A.; Sánchez-Sánchez, R.; López-Casas, P.P.; et al. Tumor Suppressor Role of RBM22 in Prostate Cancer Acting as a Dual-Factor Regulating Alternative Splicing and Transcription of Key Oncogenic Genes. *Transl. Res.* **2023**, *253*, 68–79. [CrossRef]
68. Rueden, C.T.; Schindelin, J.; Hiner, M.C.; Dezonia, B.E.; Walter, A.E.; Arena, E.T.; Eliceiri, K.W. ImageJ2: ImageJ for the next Generation of Scientific Image Data. *BMC Bioinform.* **2017**, *18*, 529. [CrossRef]
69. Watson, J.V.; Chambers, S.H.; Smith, P.J. A Pragmatic Approach to the Analysis of DNA Histograms with a Definable G1 Peak. *Cytometry* **1987**, *8*, 1–8. [CrossRef] [PubMed]
70. Vandesompele, J.; De Preter, K.; Pattyn, F.; Poppe, B.; Van Roy, N.; De Paepe, A.; Speleman, F. Accurate Normalization of Real-Time Quantitative RT-PCR Data by Geometric Averaging of Multiple Internal Control Genes. *Genome Biol.* **2002**, *3*, research0034.1. [CrossRef] [PubMed]

International Journal of
Molecular Sciences

MDPI

Article

PROM1 and EFTUD2 Expression in High-Grade Clear Cell Renal Cell Carcinoma as a Molecular Marker for Survival Rate

Michał Kasperczak [1,2,*], Iga Kołodziejczak-Guglas [2], Filip Kasperczak [1], Maciej Wiznerowicz [1,2,3] and Andrzej Antczak [1,2]

1 Department of Urology, J. Struś Hospital in Poznań, Szwajcarska 3, 61-285 Poznan, Poland; filiperkasperczako@gmail.com (F.K.); maciej.wiznerowicz@iimo.pl (M.W.); aa26@poczta.onet.pl (A.A.)
2 International Institute for Molecular Oncology, 60-203 Poznań, Poland; iga.kolodziejczak@iimo.pl
3 University Hospital of Lord's Transfiguration, Poznań University of Medical Sciences, 61-848 Poznań, Poland
* Correspondence: mkasperczakmd@gmail.com; Tel.: +48-693964628

Abstract

Clear cell renal cell carcinoma (ccRCC) is a significant global cancer, particularly impacting individuals in Western countries. Despite that, the molecular mechanisms driving renal cell carcinoma progression remain poorly understood, highlighting the need to investigate these mechanisms and identify novel therapeutic targets. Literature evidence suggests that elongation factor Tu GTP binding domain containing 2 (EFTUD2) and prominin (PROM1) gene expression have significant diagnostic potential in early tumor detection, potentially reflecting the trends in progression, and may become a novel therapeutic target. Therefore, this study aimed to evaluate EFTUD2 and PROM1 protein expression on clinical characteristics of ccRCC patients, especially overall and progression-free survival. To achieve that goal, we have combined publicly available liquid chromatography–mass spectrometry (LC-MS/MS) protein expression data with a comprehensive literature review to identify key protein markers for further study and immunohistochemical (IHC) analysis. Our findings highlight the importance of considering protein expression heterogeneity within tumors. The significant variation in EFTUD2 expression, its association with PFS, and its intricate connections with the mRNA splicing machinery underscore the need for a more nuanced understanding of its role in ccRCC. Similarly, the downregulation of PROM1 and its potential effects on cell surface interactions warrant further exploration. Future studies should focus on elucidating the mechanisms underlying these observations, exploring their potential as therapeutic targets, and investigating the specific pathways affected by their dysregulation.

Keywords: PROM1; EFTUD2; clear cell renal carcinoma; proteomics; immunohistochemistry

1. Introduction

Clear cell renal cell carcinoma (ccRCC), accounting for approximately 3% of global cancer diagnoses, exhibits a high prevalence in Western countries [1,2]. In developed nations, incidental detection via ultrasound, CT, or MRI is the primary mode of diagnosis [3]. The classic triad of hematuria, flank pain, and palpable mass, indicative of renal cell carcinomas (RCC), is observed in only 10% of patients [3]. Representing 75% of all RCCs, ccRCC originates from renal stem cells within the tubular epithelium and proximal nephron [4]. These tumors are prone to hematogenous metastasis, primarily to the liver, lungs, and bones [4]. Inactivation or deletion of the von Hippel–Lindau (VHL) tumor suppressor gene on chromosome 3p is implicated in up to 45% of ccRCC cases [5]. Germline mutations in the VHL gene contribute to 5% of cases (VHL disease), and individuals with tuberous sclerosis gene mutations exhibit an elevated risk of bilateral RCC before age 46 [5]. Modifiable risk factors for RCC include dietary habits, alcohol consumption, obesity, poorly controlled hypertension, smoking, and occupational exposures [6]. Strategies to enhance survival and mitigate disparities emphasize lifestyle modifications, improved healthcare access for underserved populations, and refined imaging guidelines for early detection [7]. Notably, 30% of ccRCC patients present with distant metastasis at diagnosis, resulting in a poor prognosis [8]. The underlying molecular mechanisms of RCC remain incompletely understood, necessitating further investigation to identify novel therapeutic targets.

EFTUD2 (Snu114) forms a stable protein complex and is a constitutive component of the U5 snRNP [9]. This protein-coding gene encoding a spliceosomal GTPase plays a pivotal role in splicing precursor mRNAs (pre-mRNAs) into mature mRNAs [10]. Recent research has revealed that elevated EFTUD2 expression in hepatocellular carcinomas is closely linked to tumor growth and poor survival [11]. This association is substantiated by the observation that EFTUD2 sustains cancer cell viability and enhances the progression of hepatocellular carcinoma through STAT3 activation [12]. Notably, EFTUD2 inhibition has been shown to attenuate the NF-KB signaling pathway's activation, leading to reduced inflammatory and tumorigenic cytokines in colon cancer [13]. The abovementioned studies imply that EFTUD2 may have additional roles in modulating cancer development, independently or dependently of its canonical function in splicing. Nonetheless, whether EFTUD2 is involved in regulating chemoresistance in different types of cancers, along with the precise molecular mechanisms of its action, remains to be understood.

In turn, prominin-1 (PROM1), also called CD133, is a membrane protein with two bulky extracellular loops with eight glycosylation residues and a ~50 amino acid-long C-terminal tail [14]. As a lipid raft protein, PROM1 is expressed mainly in microvilli or membrane protrusions of various stem cells or progenitor cells [14,15]. PROM1 is a well-known marker of cancer stem cells (CSCs) [16–18], a subpopulation of tumor cells that are thought to be drivers of tumor initiation, recurrence, and metastasis [19]. For this reason, PROM1 has been widely studied as a target for cancer therapeutics [16–18,20]. In ccRCC, PROM1 demonstrates downregulated expression [21,22]. While PROM1 contributes to cell differentiation in various tissues, its biological role remains unclear [23,24]. In RCC, PROM1 progenitor cells have been observed to differentiate into endothelial cells, promoting vascularization and tumor growth [25,26]. Additionally, its encoding product, CD133, is also downregulated in ccRCC [27]. Although PROM1 serves as a prognostic marker in other cancers like liver and ovarian cancer, its specific functional implications within ccRCC require further investigation [28,29].

This evidence suggests that EFTUD2 and PROM1 gene expression have significant diagnostic potential in early tumor detection, potentially reflecting the trends in progression. However, while previous studies have primarily focused on transcriptomic data or other cancer types, there remains a critical gap in understanding the protein-level expression of these markers in ccRCC and their association with clinical outcomes. Therefore, our study uniquely combines LC-MS/MS proteomics and immunohistochemistry to assess EFTUD2 and PROM1 protein expression in ccRCC, providing novel insights into their prognostic value.

2. Results

2.1. Protein Expression Values in CPTAC ccRCC

Among the proteins analyzed, EFTUD2 showed lower abundance in ccRCC tissues compared to NATs (Figure 1). Conversely, PROM1 exhibited lower levels in tumor samples compared to NATs. These differences were statistically significant ($p < 0.0001$ for EFTUD2 and $p = 0.0027$ for PROM1), as determined by the Mann–Whitney U test. Interestingly, we observed a substantial spread in EFTUD2 expression within tumor samples, with some specimens showing extremely high, and others extremely low, levels of this protein.

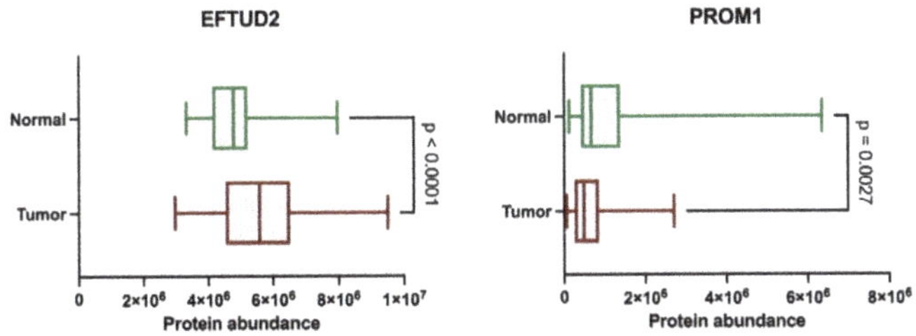

Figure 1. Distribution of protein abundance values for PROM1 (**right**) and EFTUD2 (**left**) between normal adjacent tissue ($n = 84$) and tumor ($n = 110$) samples, as measured by LC-MS/MS. y-axis represent sample groups (Normal and Tumor) and x-axis indicate protein abundance (log2-transformed intensities). Green boxes represent normal adjacent tissue samples, and red boxes denote tumor samples. For EFTUD2, $p < 0.0001$, and for PROM1, $p = 0.0027$.

2.2. Immunohistochemical Analysis of EFTUD2 and PROM1

Immunohistochemical analysis from the ccRCC validation cohort revealed that EFTUD2 expression is predominantly within the nuclei of tumor cells, while PROM1 is located in the cytoplasm (Figure 2). Many cells show some level of positivity with heterogeneous staining intensity.

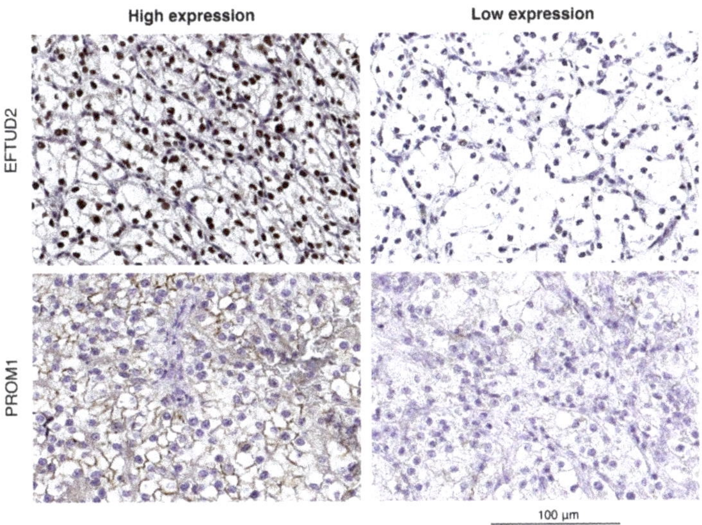

Figure 2. Immunohistochemical analysis showing immunostaining patterns in representative tumor sections for EFTUD2 and PROM1 high and low expression in the ccRCC validation cohort. Brown color indicates positive immunoreaction. Expression levels are based on staining intensity and evaluated by H-score. Scale bar = 100 μm.

2.3. EFTUD2 and PROM1 Protein Expression as Measured by IHC

For the analyzed validation cohort, samples were categorized into "high score" and "low score" groups based on their obtained IHC scores for EFTUD2 and PROM1 expression. Figure 3 provides a detailed breakdown of these expression groups. The results confirm the findings from the CPTAC protein expression analysis, demonstrating multiple instances of both very high and very low protein expression in the analyzed tumor samples. EFTUD2 generally presented higher expression in the analyzed samples, along with a significant spread of scores within the high IHC score group. The low IHC score groups exhibited a smaller spread of scores and fewer notable outliers for both proteins.

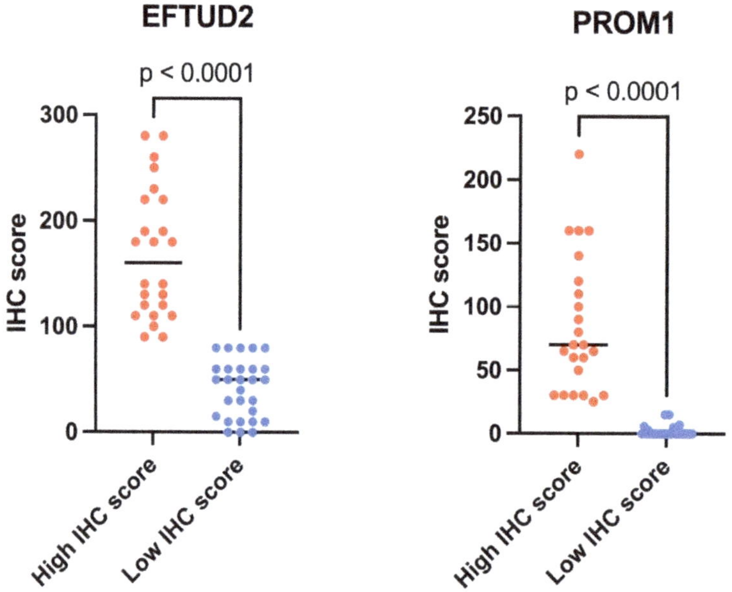

Figure 3. The detailed breakdown of the high and low EFTUD2 and PROM1 IHC score groups. The expression of both proteins shows statistically significant differences ($p < 0.0001$) between high and low score groups.

2.4. EFTUD2 and PROM1 Protein Expression Correlated with Patients' Clinical Outcomes

To assess the clinical relevance of EFTUD2 and PROM1 protein expression in ccRCC patients, we compared OS and PFS between the high and low IHC score groups. The results of the analysis are presented in Figure 4. The only significant difference ($p = 0.005$) was detected in PFS between high and low EFTUD2 expression groups. Patients with low EFTUD2 IHC scores showed worse PFS compared to patients with high EFTUD2 IHC scores. This difference was associated with a 5.016 (95% CI 1.737–14.49) Cox hazard ratio for the low EFTUD2 IHC score group.

2.5. Identification of the Top Significantly Enriched Gene Ontologies Associated with EFTUD2 and PROM1

The results of GO analysis (Figure 5) indicate a significant enrichment of ontologies related to kidney development and mRNA processing. Specifically, PROM1 is associated with several Gene Ontology terms related to podocyte and epithelial cell differentiation in the kidney. EFTUD2, on the other hand, shows a significant association with mRNA processing and splicing terms. These findings suggest distinct roles of PROM1 and EFTUD2 in kidney development and mRNA metabolism, respectively.

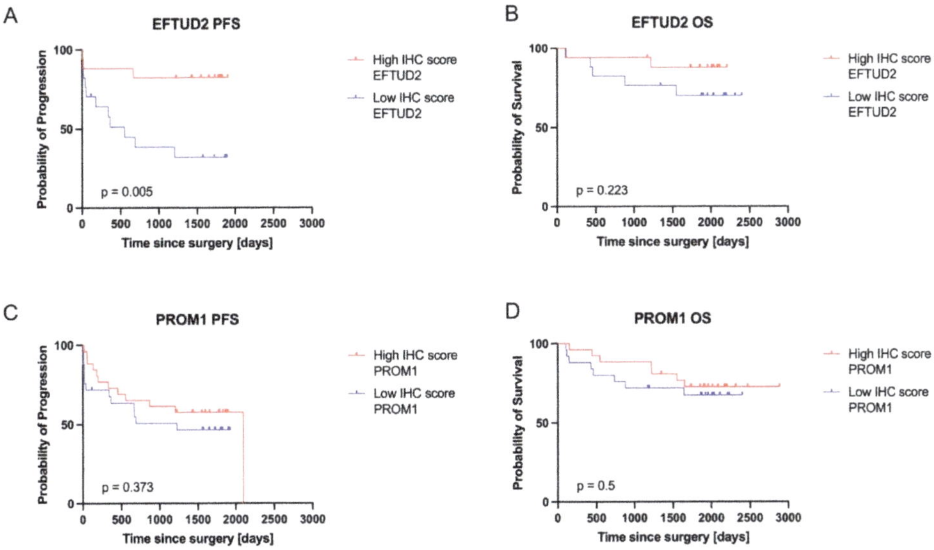

Figure 4. Kaplan–Meier survival curves comparing overall survival (OS) (**B**) between high and low EFTUD2 IHC score groups in ccRCC patients in the validation cohort; (**D**) between high and low PROM1 IHC score groups in ccRCC patients in the validation cohort; and progression-free survival (PFS) (**A**) between high and low EFTUD2 IHC score groups in ccRCC patients in the validation cohort; (**C**) between high and low PROM1 IHC score groups in ccRCC patients in the validation cohort. Groups were defined based on protein-specific H-score cutoffs calculated as the average H-score across all samples (EFTUD2 cutoff = 90; PROM1 cutoff = 25; see Section 4 Methods and Table 2). *p*-values from the log-rank test are presented in the bottom-left corner of each graph.

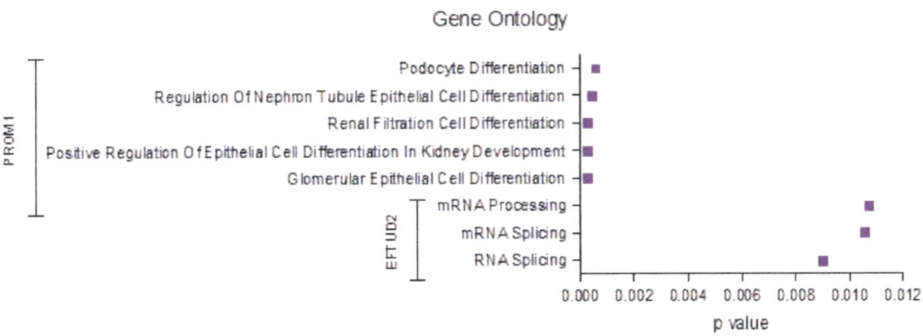

Figure 5. The results of Gene Ontology enrichment analysis for PROM1 and EFTUD2. The y-axis lists all significantly enriched GO terms, and the x-axis represents the corresponding *p*-values. Terms were identified based on statistical significance in enrichment analysis and reflect biological processes associated with each protein.

2.6. Protein–Protein Interaction Networks Reveal Key Associations of EFTUD2 and PROM1 at the Protein Level

The protein–protein interaction analysis for EFTUD2 highlights a cluster of genes primarily associated with mRNA processing and splicing (Figure 6A). The PPI network reveals associations among EFTUD2, SNRPD3, SNRPA1, SF3A1, LSM8, and SNRNP40, which have been previously implicated in mRNA processing and splicing. Furthermore, the analysis for PROM1 revealed a cluster associated with stemness and cell adhesion (Figure 6B). Genes such as PROM1, ALDH1A1, EPCAM, SOX2, NANOG, and POU5F1 form a highly interconnected cluster in the PPI network and are commonly associated with stem cell maintenance and differentiation. Additionally, the presence of genes like CDH1, CD24, CD44, and CD34 may suggest involvement in cell adhesion and signaling within this network.

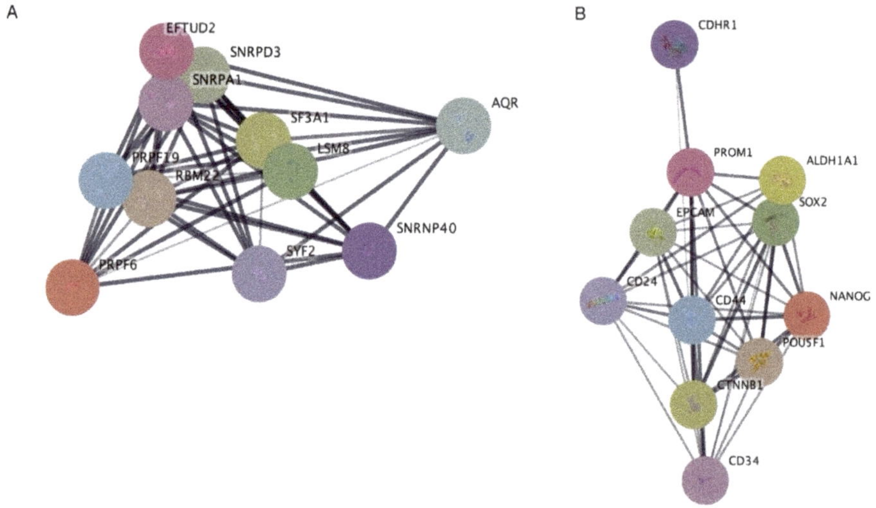

Figure 6. The results of protein–protein interaction (PPI) analysis for EFTUD2 (**A**) and PROM1 (**B**), performed in Cytoscape based on STRING database data. Each node (colored ball) represents a protein, with a 3D structural preview displayed inside the node. Edges (connecting lines) represent predicted or known protein–protein interactions.

3. Discussion

This study investigated the expression of EFTUD2 and PROM1 in ccRCC and their potential as prognostic markers. Our analysis of the CPTAC ccRCC dataset revealed significant differences in the abundance of these proteins between tumor tissues and NATs. EFTUD2 was significantly upregulated in ccRCC, suggesting a potential involvement in tumor progression, while PROM1 was downregulated. The wide range of EFTUD2 expression in tumor samples, from extremely high to extremely low, was further corroborated by IHC analysis, indicating heterogeneity within ccRCC.

IHC staining provided further insights into the cellular localization of these proteins. Specifically, we found that EFTUD2 expression is predominantly within the nuclei of tumor cells. This localization is consistent with our analysis showing that EFTUD2 interacts with numerous nuclear proteins involved in mRNA splicing, including core spliceosome components (SNRPD3, SNRPA1), proteins essential for snRNP formation (SF3A1, LSM8, SNRNP40), and factors involved in spliceosome assembly, catalysis, and regulation (PRPF19, RBM22, AQR, PRPF6). Notably, EFTUD2 also interacts with SYF2, linking cell cycle regulation to splicing. This diverse interactome underscores the complexity of EFTUD2's role in RNA processing, suggesting that its expression levels could influence splicing fidelity and efficiency, with potential downstream effects on gene expression and cellular function [30].

Our GO enrichment analysis further suggests its involvement in mRNA splicing. It is plausible that disruptions in EFTUD2-mediated splicing contribute to the dysregulation of gene expression programs critical for tumor suppression, leading to a more aggressive phenotype and poorer clinical outcomes. Crucially, correlating protein expression with clinical outcomes revealed that low EFTUD2 IHC scores were significantly associated with worse PFS in ccRCC patients. This is particularly interesting given the observed variability in EFTUD2 expression within tumor samples, suggesting that while EFTUD2 is generally upregulated in tumors, its reduced expression might indicate a more aggressive disease. This hypothesis is supported by studies in other cancers, such as colorectal cancer, where EFTUD2 has been shown to influence c-MYC stability and contribute to chemoresistance [31].

Beyond its role in mRNA splicing, EFTUD2 appears to have broader functions in cancer development. It has been implicated in modulating NF-κB signaling [31], influencing cell viability and migration [32], and regulating the innate macrophage immune response [33], ultimately serving as a potential biomarker in different cancer types [10,34]. These diverse roles highlight the complex and context-dependent nature of EFTUD2 in cancer. For example, in endometrial cancer, EFTUD2 expression correlated with poor prognosis and was linked to increased expression of RIG-I, suggesting an interplay between EFTUD2 and the tumor microenvironment [35]. Furthermore, research in hepatocellular carcinoma has revealed a connection between EFTUD2 and metabolic reprogramming, where EFTUD2 regulates the protein stability of YTHDF3, an m6A "reader" protein involved in RNA metabolism [36]. Finally, EFTUD2 may be used as a therapeutic strategy, as identified by Sato et al., where its knockdown in breast cancer cells led to an increase in apoptotic cells [37].

PROM1 (CD133) was primarily found on the cell membrane, in line with its role as a transmembrane stem cell marker [38]. This localization allows PROM1 to interact with various cell surface proteins involved in diverse cellular processes. Our results show that PROM1 interacts with cadherins like CDHR1 and cell adhesion molecules like EpCAM, potentially influencing cell–cell adhesion and tissue integrity. Its interaction with ALDH1A1 could influence cellular differentiation and response to oxidative stress. Furthermore, it interacts with key stem cell regulators (e.g., SOX2, NANOG, POU5F1), suggesting a role in stem cell maintenance and differentiation, and with molecules like CD24, CD44, and CTNNB1, potentially affecting cell adhesion, migration, and signaling. The interaction with CD34, a hematopoietic stem cell marker, could be relevant in the context of cancer stem cells (CSC). These interactions highlight the complex role of PROM1 in cellular communication and function. Indeed, our GO enrichment analysis further showed that PROM1 appears to be associated with kidney development and epithelial cell differentiation, supporting its multifaceted involvement in both normal tissue development and potentially aberrant processes in cancer.

While our study found it to be downregulated in ccRCC compared to NAT without showing any significant correlation to the patients' clinical condition, some studies have associated high PROM1 expression with poor prognosis, [39,40]. This discrepancy might be explained by the diverse roles PROM1 plays in different cellular contexts. For instance, PROM1 can be expressed on both stem cells and differentiated cells, and its glycosylation status can vary with cellular differentiation and malignant transformation [41–43]. Further highlighting PROM1's complex role, different studies reveal that PROM1's prognostic significance varies across different cancer types, with its expression and methylation having contrasting implications in gliomas and papillary RCC [44]. Additionally, research in thyroid cancer has shown that its expression varies across different subtypes and that a specific gene signature, including high PROM1 expression, correlates with worse recurrence-free survival [27]. Furthermore, the localization of PROM1 within the cell may influence its function, with cytoplasmic expression potentially being associated with increased tumor aggressiveness [45]. This complexity underscores the challenges in identifying and characterizing CSCs in ccRCC, as single marker expression may not accurately reflect the CSC population [46]. Indeed, some studies suggest that PROM1+ cells in ccRCC may represent a heterogeneous population [43]. This heterogeneity has significant implications for understanding ccRCC progression and treatment response. If PROM1+ cells have varying degrees of stemness and differentiation, therapies targeting PROM1 may have unpredictable effects.

Despite these challenges, PROM1 holds potential as a biomarker in ccRCC. Studies have explored its use in predicting metastasis and recurrence, warranting further investigation [27]. Interestingly, targeting specific pathways in PROM1+ CSCs can influence their survival and sensitivity to therapy. For example, activating TNFR2 signaling promotes the proliferation of PROM1+ CSCs in ccRCC and increases their sensitivity to cell cycle-dependent cytotoxicity [47]. In line with the complex role of PROM1 in ccRCC, its involvement in other cancers also appears to be context dependent. For example, in ovarian cancer, PROM1 overexpression has been linked to poor prognosis and worse response to treatment, particularly in patients with TP53 mutations [48].

These findings, along with our observation of EFTUD2's association with PFS in ccRCC, emphasize the importance of considering these proteins' diverse roles and interactions in cancer development. Our study contributes to a growing body of knowledge highlighting the importance of protein–protein interaction and network perturbations in tumor progression [49], underscoring the need for further investigation into the complex interplay of these factors in ccRCC. However, interpretations of the results should be cautious, as the study's modest sample size and focused design, while intentional for exploratory comparisons, suggest that these findings are preliminary and necessitate validation in larger, independent cohorts. The absence of formal multiple testing correction further underscores the preliminary nature of these findings and the need for validation in larger studies employing appropriate statistical methods.

This work further underscores the emerging significance of non-traditional biomarkers and molecular features in predicting RCC outcomes. Unlike conventional clinical or imaging parameters, which often provide a static snapshot, these molecular indicators offer deeper, more dynamic insights into tumor biology. Our investigation into the protein expression patterns and interactions of EFTUD2 and PROM1 aligns with this evolving landscape, where understanding the nuanced molecular underpinnings of ccRCC is paramount. These non-traditional markers encompass a diverse range of molecular characteristics, such as specific protein expression patterns, post-translational modifications, and intricate protein–protein interaction networks within tissues [50–52]. Moreover, important examples also include those detectable in easily accessible biological samples like blood or urine, such as circulating tumor DNA (ctDNA), circulating tumor cells (CTCs), and microRNAs (miRNAs) [29,53,54]. The integration of these advanced molecular insights is crucial for moving beyond broad disease classifications to achieve more precise prognostic and predictive models for individual patients [55]. By capturing real-time disease progression and tumor heterogeneity, these markers can significantly refine risk stratification and guide personalized therapeutic approaches, ultimately aiming to improve clinical decision-making and patient outcomes [56]. Furthermore, considering the significant clinical relevance of metastasectomy and its impact on survival, particularly in carefully selected patients with oligometastatic disease where complete surgical removal of metastatic lesions can lead to prolonged survival in renal cell carcinoma, some reflection in this direction would situate the study in a broader clinical context [57,58]. While our study focuses on initial prognostic markers, the utility of such markers could extend to informing patient selection for metastasectomy, potentially refining patient management strategies and improving overall survival in those with oligometastatic disease [58].

Our study has several limitations that should be acknowledged. Firstly, our analysis is primarily based on correlational data. Further mechanistic studies are needed to fully substantiate our findings and elucidate the precise molecular mechanisms by which EFTUD2 and PROM1 influence ccRCC progression. Secondly, our study focused on profiling EFTUD2 and PROM1 expression, limiting the scope of our investigation. We did not include a comprehensive analysis of clinical/histological progression, a critical aspect of ccRCC, nor did we examine the role of these proteins in the context of therapies such as chemotherapy, which is relevant for advanced-stage disease. While we leveraged CPTAC's proteomic data, future studies could integrate its genomic information for deeper mechanistic insights, expanding upon our current findings. Additionally, due to our limited sample size, we could not perform stratified analyses by variables like histological grade or age group. While such analyses would offer valuable subgroup-specific insights into metastatic risk, our current cohort is underpowered for this. Additionally, our study lacks in vitro or in vivo experimental validation, which will be essential to confirm the functional relevance of our findings in a biological context. Future studies should investigate the complex interplay between EFTUD2/PROM1 expression, clinical/histological progression, and therapeutic responses in ccRCC to provide a more complete understanding of their roles in this disease, and larger cohorts are needed to enable robust stratification.

4. Materials and Methods

4.1. Patients and Cohorts

Our study utilized a ccRCC discovery cohort from the Clinical Proteomic Tumor Analysis Consortium (CPTAC), which included 110 untreated ccRCC cases and 84 matched normal adjacent tissue (NAT) samples, as described by Clark et al. [30]. We integrated publicly available LC-MS/MS protein expression data (referred to as protein abundance, which are log2-transformed intensities) with a comprehensive literature review to identify key protein markers for subsequent investigation and immunohistochemical (IHC) analysis (Figure 7).

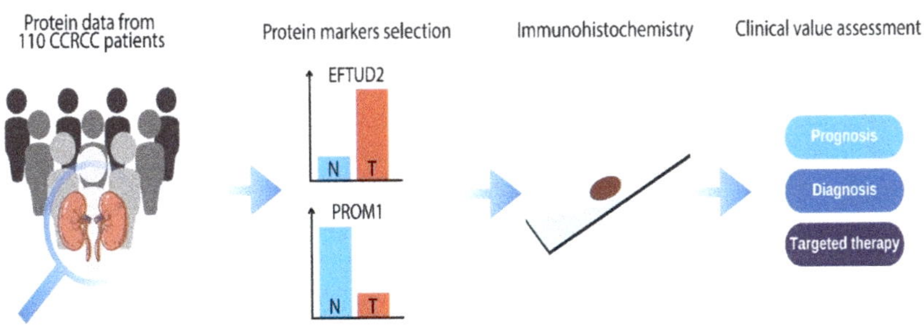

Figure 7. Study design overview: The study involved the acquisition of protein data from the CPTAC discovery ccRCC cohort dataset and subsequent analysis of protein expression of tumor (T, $n = 110$) and normal adjacent tissue (N, $n = 84$) to select candidate protein markers. The selected protein markers were then used as a target of immunohistochemical analysis. Finally, we have conducted an assessment of the clinical value of the selected proteins for prognosis, diagnosis, and targeted therapy for ccRCC patients.

The proteomic data from the CPTAC dataset allowed for a thorough examination of protein markers involved in ccRCC development. Based on their biological relevance, we selected proteins exhibiting differential expression in tumors compared to normal adjacent tissues: EFTUD2 (higher expression) and PROM1 (lower expression). The expression of these validated markers was then linked to clinical data via IHC analysis. This integrated approach, combining proteomic data acquisition, marker validation, and clinical correlation, enhances our understanding of ccRCC biology, progression, and potential clinical implications.

The validation cohort consisted of 52 ccRCC samples from individuals aged 31 to 84 years, with 23 of these participants also part of the CPTAC discovery study. Detailed demographic and clinical information, including age, gender, race, grade, and stage, was collected. Over a five-year follow-up period, 26 patients developed metastasis, while 26 did not. This study focused on adult patients with histopathologically confirmed ccRCC tumors. Ethical approval was obtained in accordance with CPTAC guidelines. Patients who had received systemic treatment or were diagnosed with other cancers in the previous 12 months were excluded from the study.

4.2. Immunohistochemistry and Pathology Evaluation

IHC staining was performed using the Dako Autostainer Link 48, with preprogrammed staining protocols and the EnVision visualizing kit (Cat No. K800221-2, Dako, Agilent Technologies Inc., Carpinteria, CA, USA). This polymer-based system directly conjugates HRP to the secondary antibody complex, eliminating the need for a separate secondary antibody. The reaction was visualized using 3,3′-diaminobenzidine (DAB) as the chromogen.

Tissue samples were collected retrospectively, fixed in 10% neutral buffered formalin for 24 h, and then embedded in paraffin. Following this, 4-micron sections were cut from these formalin-fixed paraffin-embedded (FFPE) blocks and placed on positively charged slides. Antigen retrieval was carried out using Dako Target Retrieval Solution, High pH (Cat No. S2367), in a PT Link Pre-Treatment Module (Dako, Agilent Technologies Inc.) at 97 °C for 20 min. After antigen retrieval, sections were incubated with primary antibodies for 30 min at room temperature. Two primary antibodies were used for the clinical validation cohort, as detailed in Table 1.

Table 1. Primary antibodies used for immunohistochemistry in this study.

Antibody	Source	Identifier	Clone	Dilution
Rabbit polyclonal anti-EFTUD2	Atlas Antibodies	Cat# HPA022021, RRID:AB_1848032	Polyclonal	1:400
Rabbit polyclonal anti-PROM1	Atlas Antibodies	Cat# HPA004922, RRID:AB_1846238	Polyclonal	1:250

The IHC-stained slides were independently assessed by three pathologists using the H-score method. The H-score is a semi-quantitative method that considers both staining intensity and the percentage of positive cells, resulting in a score from 0 to 300. The equation used for H-score calculation is as follows:

$$\text{H-score} = (\% \text{ of cells stained at intensity } 1 \times 1) + (\% \text{ of cells stained at intensity } 2 \times 2) + (\% \text{ of cells stained at intensity } 3 \times 3)$$

The elements of this formula are represented as follows:

- Intensity 1 indicates weak staining;
- Intensity 2 indicates moderate staining;
- Intensity 3 indicates strong staining.

These percentages represent the proportion of cells at each staining intensity. The H-score, derived by summing the products of intensity scores and their corresponding percentages, provides a comprehensive assessment of both the quantity and intensity of IHC staining in a specimen [31].

A total of 52 ccRCC FFPE samples were used to evaluate PROM1 and EFTUD2 expression through IHC analysis. These samples were categorized into groups for further analysis based on their H-scores. Patients with IHC scores above the cutoff value were placed in the "High IHC score" group, while those with scores below the cutoff were placed in the "Low IHC score" group (Table 2). The cutoff value for each protein was determined by calculating the average of all H-scores from the 52 IHC slides. For EFTUD2, the cutoff was set at an H-score of 90; samples with H-scores ≥ 90 were assigned to the High IHC score group. For PROM1, the cutoff was established at an H-score of 25, with scores $\geq$ categorized as High IHC expression.

Table 2. Classification of IHC specimen based on H-scores.

Protein	H-Score Range—High IHC Score Group	H-Score Range—Low IHC Score Group	Reaction
EFTUD2	90–280, $n = 24$	10–80, $n = 28$	Nuclear
PROM1	25–220, $n = 24$	0–15, $n = 28$	Cytoplasmic, membranous

4.3. Digital Image Acquisition and Archiving

Digital scans of all IHC slides were generated at 20× magnification using a ScanScope AT Turbo whole slide scanner (Aperio/Leica Microsystems, Vista, CA, USA). These digital images, saved in .svs format, were subsequently reviewed by expert pathologists using ImageScope software (version 12.3.3, Aperio, Vista, CA, USA) for detailed assessment. Access to the digitized images was provided via a password-protected Synology Rack Station server (RS18017xs+).

4.4. Statistical Analysis

Differences in protein abundance between normal and tumor tissues in ccRCC were computed using a non-parametric Wilcoxon rank-sum test. For comparing IHC scores between the High IHC score and Low IHC score groups of the validation cohort, the Wilcoxon rank-sum test was also performed. Additionally, differences in protein abundance between groups were assessed using the Mann–Whitney U test.

Overall survival (OS) and progression-free survival (PFS) analyses for the validation cohort were conducted by dividing the validation cohort into two groups (High IHC score and Low IHC score, separately for each protein target). These groups were then compared using the Kaplan–Meier method and the log-rank (Mantel–Cox) test. Hazard ratios (HR) and their 95% confidence intervals (CI) were calculated to assess the impact on survival outcomes. All data were analyzed using GraphPad Prism 10 software. A p-value < 0.05 was considered statistically significant for all analyses.

4.5. Gene Ontology Enrichment Analysis

To gain deeper insights into the biological processes and tumor mechanisms associated with PROM1 and EFTUD2, we conducted a Gene Ontology (GO) enrichment analysis.

This analysis aimed to identify the most significant Reactome pathways enriched for each protein, thereby elucidating their functional implications in ccRCC.

Gene Ontology enrichment analysis for EFTUD2 and PROM1 was conducted using the Enrichr platform (available at https://maayanlab.cloud/Enrichr (accessed on 13 April 2025)), a comprehensive and interactive web-based tool for gene list enrichment analysis. GO terms were evaluated across the Biological Process (BP) category. Enrichr computes enrichment by combining statistical significance from Fisher's exact test with a correction for multiple hypothesis testing. The output includes GO terms ranked by adjusted p-values and combined scores, enabling identification of overrepresented biological annotations associated with the input genes. A statistical significance threshold of $p < 0.05$ was applied to determine the biological processes, ensuring that the identified processes were highly likely to be biologically relevant [32–34].

4.6. Protein–Protein Interaction as Determined by String Database

Protein–protein interaction (PPI) networks for EFTUD2 and PROM1 were constructed using the stringApp for STRING database (version 2.1.1) integrated with Cytoscape (version 3.10). Individual analyses were conducted for each protein to identify their interaction protein partners. STRING was queried with the official gene symbols "EFTUD2" and "PROM1," and interactions were retrieved based on high-confidence scores (e.g., confidence score ≥ 0.8). The network construction included both direct (physical) and indirect (functional) associations, limited to interactions supported by experimental evidence or curated databases. The resulting PPI networks were visualized and analyzed in Cytoscape, showing only statistically significant results ($p < 0.05$). Node sizes and edge thickness were adjusted to reflect the number of interactions and confidence levels, respectively.

5. Conclusions

Our findings highlight the importance of considering protein expression heterogeneity within tumors. The significant variation in EFTUD2 expression, its association with PFS, and its intricate connections with the mRNA splicing machinery underscore the need for a more nuanced understanding of its role in ccRCC. Similarly, the downregulation of PROM1 and its potential effects on cell surface interactions warrant further exploration. Future studies should focus on elucidating the mechanisms underlying these observations, exploring their potential as therapeutic targets, and investigating the specific pathways affected by their dysregulation.

Author Contributions: Conceptualization, M.K. and I.K.-G.; methodology, F.K.; software, F.K.; validation, M.K. and I.K.-G.; formal analysis, M.K. and I.K.-G.; investigation, M.K. and I.K.-G.; resources, I.K.-G.; data curation, M.K. and I.K.-G.; writing—original draft preparation, M.K.; writing—review and editing, M.K. and I.K.-G.; visualization, M.K.; supervision, A.A. and M.W.; project administration, M.W.; funding acquisition, M.K. All authors have read and agreed to the published version of the manuscript.

Funding: This research received no external funding.

Institutional Review Board Statement: The study was conducted in accordance with the Declaration of Helsinki and approved by the Institutional Ethics Committee of Poznań University of Medical Sciences KB: 302/21. Approval Date: 8 April 2021.

Informed Consent Statement: Informed consent was obtained from all subjects involved in the study.

Data Availability Statement: All data on which the manuscript was based, if not already included in its text, are available from the corresponding author upon reasonable request.

Conflicts of Interest: The authors declare no conflicts of interest.

References

1. Ferlay, J.; Colombet, M.; Soerjomataram, I.; Dyba, T.; Randi, G.; Bettio, M.; Gavin, A.; Visser, O.; Bray, F. Cancer incidence and mortality patterns in Europe: Estimates for 40 countries and 25 major cancers in 2018. *Eur. J. Cancer* **2018**, *103*, 356–387. [CrossRef] [PubMed]
2. Capitanio, U.; Bensalah, K.; Bex, A.; Boorjian, S.A.; Bray, F.; Coleman, J.; Gore, J.L.; Sun, M.; Wood, C.; Russo, P.; et al. Epidemiology of Renal Cell Carcinoma. *Eur. Urol.* **2019**, *75*, 74–84. [CrossRef] [PubMed]
3. DeCastro, G.J.; McKiernan, J.M. Epidemiology, Clinical Staging, and Presentation of Renal Cell Carcinoma. *Urol. Clin. N. Am.* **2008**, *35*, 581–592. [CrossRef] [PubMed]
4. Chevrier, S.; Levine, J.H.; Zanotelli, V.R.T.; Silina, K.; Schulz, D.; Bacac, M.; Ries, C.H.; Ailles, L.; Jewett, M.A.S.; Moch, H.; et al. An Immune Atlas of Clear Cell Renal Cell Carcinoma. *Cell* **2017**, *169*, 736–749.e18. [CrossRef]
5. Creighton, C.J.; Hernandez-Herrera, A.; Jacobsen, A.; Levine, D.A.; Mankoo, P.; Schultz, N.; Du, Y.; Zhang, Y.; Larsson, E.; Sheridan, R.; et al. Integrated analyses of microRNAs demonstrate their widespread influence on gene expression in high-grade serous ovarian carcinoma. *PLoS ONE* **2012**, *7*, e34546. [CrossRef]
6. Chow, W.H.; Dong, L.M.; Devesa, S.S. Epidemiology and risk factors for kidney cancer. *Urology* **2010**, *7*, 245–257. [CrossRef]
7. Howlader, N.; Noone, A.M.; Krapcho, M.; Miller, D.; Brest, A.; Yu, M.; Ruhl, J.; Tatalovich, Z.; Mariotto, A.; Lewis, D.R.; et al. *SEER Cancer Statistics Review, 1975–2016*; National Cancer Institute: Rockville, MD, USA, 2019.
8. Capitanio, U.; Montorsi, F. Renal cancer. *Lancet* **2016**, *387*, 894–906. [CrossRef]
9. Achsel, T.; Ahrens, K.; Brahms, H.; Teigelkamp, S.; Lührmann, R. The Human U5-220kD Protein (hPrp8) Forms a Stable RNA-Free Complex with Several U5-Specific Proteins, Including an RNA Unwindase, a Homologue of Ribosomal Elongation Factor EF-2, and a Novel WD-40 Protein. *Mol. Cell Biol.* **1998**, *18*, 6756–6766. [CrossRef]
10. Wood, K.A.; Eadsforth, M.A.; Newman, W.G.; O'Keefe, R.T. The Role of the U5 snRNP in Genetic Disorders and Cancer. *Front. Genet.* **2021**, *12*, 636620. [CrossRef]
11. Lv, C.; Li, X.J.; Hao, L.X.; Zhang, S.; Song, Z.; Ji, X.D.; Gong, B. Over-activation of EFTUD2 correlates with tumor propagation and poor survival outcomes in hepatocellular carcinoma. *Clin. Transl. Oncol.* **2022**, *24*, 93–103. [CrossRef]
12. Tu, M.; He, L.; You, Y.; Li, J.; Yao, N.; Qu, C.; Huang, W.; Xu, L.; Luo, R.; Hong, J. EFTUD2 maintains the survival of tumor cells and promotes hepatocellular carcinoma progression via the activation of STAT3. *Cell Death Dis.* **2020**, *11*, 830. [CrossRef] [PubMed]
13. Lv, Z.; Wang, Z.; Luo, L.; Chen, Y.; Han, G.; Wang, R.; Xiao, H.; Li, X.; Hou, C.; Feng, J.; et al. Spliceosome protein Eftud2 promotes colitis-associated tumorigenesis by modulating inflammatory response of macrophage. *Mucosal Immunol* **2019**, *12*, 1164–1173. [CrossRef] [PubMed]
14. Weigmann, A.; Corbeil, D.; Hellwig, A.; Huttner, W.B. Prominin, a novel microvilli-specific polytopic membrane protein of the apical surface of epithelial cells, is targeted to plasmalemmal protrusions of non-epithelial cells. *Proc. Natl. Acad. Sci. USA* **1997**, *94*, 12425–12430. [CrossRef] [PubMed]
15. Röper, K.; Corbeil, D.; Huttner, W.B. Retention of prominin in microvilli reveals distinct cholesterol-based lipid micro-domains in the apical plasma membrane. *Nat. Cell Biol.* **2000**, *2*, 582–592. [CrossRef]
16. Liu, G.; Yuan, X.; Zeng, Z.; Tunici, P.; Ng, H.; Abdulkadir, I.R.; Lu, L.; Irvin, D.; Black, K.L.; Yu, J.S. Analysis of gene expression and chemoresistance of CD133+ cancer stem cells in glioblastoma. *Mol. Cancer* **2006**, *5*, 67. [CrossRef]
17. Bao Bao, S.; Wu, Q.; McLendon, R.E.; Hao, Y.; Shi, Q.; Hjelmeland, A.B.; Dewhirst, M.W.; Bigner, D.D.; Rich, J.N. Glioma stem cells promote radioresistance by preferential activation of the DNA damage response. *Nature* **2006**, *444*, 756–760. [CrossRef]
18. Florek, M.; Haase, M.; Marzesco, A.M.; Freund, D.; Ehninger, G.; Huttner, W.B.; Corbeil, D. Prominin-1/CD133, a neural and hematopoietic stem cell marker, is expressed in adult human differentiated cells and certain types of kidney cancer. *Cell Tissue Res.* **2005**, *319*, 15–26. [CrossRef]
19. Ricci-Vitiani, L.; Lombardi, D.G.; Pilozzi, E.; Biffoni, M.; Todaro, M.; Peschle, C.; De Maria, R. Identification and expansion of human colon-cancer-initiating cells. *Nature* **2007**, *445*, 111–115. [CrossRef]
20. Yin, S.; Li, J.; Hu, C.; Chen, X.; Yao, M.; Yan, M.; Jiang, G.; Ge, C.; Xie, H.; Wan, D.; et al. CD133 positive hepatocellular carcinoma cells possess high capacity for tumorigenicity. *Int. J. Cancer* **2007**, *120*, 1444–1450. [CrossRef]
21. Wang, J.; Yuan, L.; Liu, X.; Wang, G.; Zhu, Y.; Qian, K.; Xiao, Y.; Wang, X. Bioinformatics and functional analyses of key genes and pathways in human clear cell renal cell carcinoma. *Oncol. Lett.* **2018**, *15*, 9133–9141. [CrossRef]

22. D'Alterio, C.; Cindolo, L.; Portella, L.; Polimeno, M.; Consales, C.; Riccio, A.; Cioffi, M.; Franco, R.; Chiodini, P.; Cartenì, G.; et al. Differential role of CD133 and CXCR4 in renal cell carcinoma. *Cell Cycle* **2010**, *9*, 4492–4500. [CrossRef] [PubMed]
23. Sagrinati, C.; Netti, G.S.; Mazzinghi, B.; Lazzeri, E.; Liotta, F.; Frosali, F.; Ronconi, E.; Meini, C.; Gacci, M.; Squecco, R.; et al. Isolation and characterization of multipotent progenitor cells from the Bowman's capsule of adult human kidneys. *J. Am. Soc. Nephrol.* **2006**, *17*, 2443–2456. [CrossRef] [PubMed]
24. Ronconi, E.; Sagrinati, C.; Angelotti, M.L.; Lazzeri, E.; Mazzinghi, B.; Ballerini, L.; Parente, E.; Becherucci, F.; Gacci, M.; Carini, M.; et al. Regeneration of glomerular podocytes by human renal progenitors. *J. Am. Soc. Nephrol.* **2009**, *20*, 322–332. [CrossRef] [PubMed]
25. Bussolati, B.; Bruno, S.; Grange, C.; Buttiglieri, S.; Deregibus, M.C.; Cantino, D.; Camussi, G. Isolation of Renal Progenitor Cells from Adult Human Kidney. *Am. J. Pathol.* **2005**, *166*, 545–555. [CrossRef]
26. Bussolati, B.; Grange, C.; Collino, F.; Graziano, M.E.; Ferrando, U.; Camussi, G. CD133+ Renal Progenitor Cells Contribute to Tumor Angiogenesis. *Am. J. Pathol.* **2006**, *169*, 2223–2235. [CrossRef]
27. Kim, K.; Ihm, H.; Ro, J.Y.; Cho, Y.M. High-level expression of stem cell marker CD133 in clear cell renal cell carcinoma with favorable prognosis. *Oncol. Lett.* **2011**, *2*, 1095. [CrossRef]
28. Zhou, L.; Yu, K.H.; Wong, T.L.; Zhang, Z.; Chan, C.H.; Loong, J.H.; Che, N.; Yu, H.J.; Tan, K.V.; Tong, M.; et al. Lineage tracing and single-cell analysis reveal proliferative Prom1+ tumour-propagating cells and their dynamic cellular transition during liver cancer progression. *Gut* **2022**, *71*, 1656–1668. [CrossRef]
29. Dansonka-Mieszkowska, A.; Szafron, L.A.; Kulesza, M.; Stachurska, A.; Leszczynski, P.; Tomczyk-Szatkowska, A.; Sobiczewski, P.; Parada, J.; Kulinczak, M.; Moes-Sosnowska, J.; et al. PROM1, CXCL8, RUNX1, NAV1 and TP73 genes as independent markers predictive of prognosis or response to treatment in two cohorts of high-grade serous ovarian cancer patients. *PLoS ONE* **2022**, *17*, e0271539. [CrossRef]
30. Clark, D.J.; Dhanasekaran, S.M.; Petralia, F.; Pan, J.; Song, X.; Hu, Y.; da Veiga Leprevost, F.; Reva, B.; Lih, T.-S.M.; Chang, H.-Y.; et al. Integrated Proteogenomic Characterization of Clear Cell Renal Cell Carcinoma. *Cell* **2019**, *179*, 964–983.e31. [CrossRef]
31. Hirsch, F.R.; Varella-Garcia, M.; Bunn, P.A., Jr.; Di Maria, M.V.; Veve, R.; Bremmes, R.M.; Barón, A.E.; Zeng, C.; Franklin, W.A. Epidermal growth factor receptor in non-small-cell lung carcinomas: Correlation between gene copy number and protein expression and impact on prognosis. *J. Clin. Oncol.* **2003**, *21*, 3798–3807. [CrossRef]
32. Xie, Z.; Bailey, A.; Kuleshov, M.V.; Clarke, D.J.B.; Evangelista, J.E.; Jenkins, S.L.; Lachmann, A.; Wojciechowicz, M.L.; Kropiwnicki, E.; Jagodnik, K.M.; et al. Gene Set Knowledge Discovery with Enrichr. *Curr. Protoc.* **2021**, *1*, e90. [CrossRef] [PubMed]
33. Chen, E.Y.; Tan, C.M.; Kou, Y.; Duan, Q.; Wang, Z.; Meirelles, G.V.; Clark, N.R.; Ma'ayan, A. Enrichr: Interactive and collaborative HTML5 gene list enrichment analysis tool. *BMC Bioinform.* **2013**, *14*, 128. [CrossRef] [PubMed]
34. Kuleshov, M.V.; Jones, M.R.; Rouillard, A.D.; Fernandez, N.F.; Duan, Q.; Wang, Z.; Koplev, S.; Jenkins, S.L.; Jagodnik, K.M.; Lachmann, A.; et al. Enrichr: A comprehensive gene set enrichment analysis web server 2016 update. *Nucleic Acids Res.* **2016**, *44*, W90–W97. [CrossRef] [PubMed]
35. Zhu, X.; Li, C.; Gao, Y.; Zhang, Q.; Wang, T.; Zhou, H.; Bu, F.; Chen, J.; Mao, X.; He, Y.; et al. The feedback loop of EFTUD2/c-MYC impedes chemotherapeutic efficacy by enhancing EFTUD2 transcription and stabilizing c-MYC protein in colorectal cancer. *J. Exp. Clin. Cancer Res.* **2024**, *43*, 7. [CrossRef]
36. Ma, H.; Suo, L.; Zhao, J.; Ma, R.; Wang, Q.; Liu, J.; Qiao, J.; Wu, J.; An, J.; Liu, Y.; et al. Prognostic biomarkers based on GUF1, EFTUD2 and GSPT1 targets affecting migration of gastric cancer cells. *Transl. Cancer Res.* **2024**, *13*, 4827–4845. [CrossRef]
37. Chen, F.; Wang, Q.; Zhou, Y. The construction and validation of an RNA binding protein-related prognostic model for bladder cancer. *BMC Cancer* **2021**, *21*, 244. [CrossRef]
38. Zhou, W.; Chen, Y.; Luo, R.; Li, Z.; Jiang, G.; Ou, X. Identification of Biomarkers Related to Immune Cell Infiltration in Hepatocellular Carcinoma Using Gene Co-Expression Network. *Pathol. Oncol. Res.* **2021**, *27*, 601693. [CrossRef]
39. Zhang, Z.G.; Shi, Z.D.; Dong, J.J.; Chen, Y.A.; Cao, M.Y.; Li, Y.T.; Ma, W.M.; Hao, L.; Pang, K.; Zhou, J.H.; et al. Novel potential urinary biomarkers for effective diagnosis and prognostic evaluation of high-grade bladder cancer. *Transl. Cancer Res.* **2023**, *12*, 1992–2007. [CrossRef]
40. Beyer, S.; Müller, L.; Mitter, S.; Keilmann, L.; Meister, S.; Buschmann, C.; Kraus, F.; Topalov, N.E.; Czogalla, B.; Trillsch, F.; et al. High RIG-I and EFTUD2 expression predicts poor survival in endometrial cancer. *J. Cancer Res. Clin. Oncol.* **2022**, *149*, 4293. [CrossRef]
41. Zhou, R.; Ni, W.; Qin, C.; Zhou, Y.; Li, Y.; Huo, J.; Bian, L.; Zhou, A.; Li, J. A functional loop between YTH domain family protein YTHDF3 mediated m6A modification and phosphofructokinase PFKL in glycolysis of hepatocellular carcinoma. *J. Exp. Clin. Cancer Res.* **2022**, *41*, 334. [CrossRef]

42. Sato, N.; Maeda, M.; Sugiyama, M.; Ito, S.; Hyodo, T.; Masuda, A.; Tsunoda, N.; Kokuryo, T.; Hamaguchi, M.; Nagino, M.; et al. Inhibition of SNW1 association with spliceosomal proteins promotes apoptosis in breast cancer cells. *Cancer Med.* **2015**, *4*, 268–277. [CrossRef] [PubMed]
43. Bahn, M.S.; Ko, Y.G. PROM1-mediated cell signal transduction in cancer stem cells and hepatocytes. *BMB Rep.* **2023**, *56*, 65–70. [CrossRef] [PubMed]
44. da Costa, W.H.; Rocha, R.M.; da Cunha, I.W.; da Fonseca, F.P.; Guimaraes, G.C.; de Cassio Zequi, S. CD133 immunohistochemical expression predicts progression and cancer-related death in renal cell carcinoma. *World J. Urol.* **2012**, *30*, 553–558. [CrossRef] [PubMed]
45. Yin, A.H.; Miraglia, S.; Zanjani, E.D.; Almeida-Porada, G.; Ogawa, M.; Leary, A.G.; Olweus, J.; Kearney, J.; Buck, D.W. AC133, a Novel Marker for Human Hematopoietic Stem and Progenitor Cells. *Blood* **1997**, *90*, 5002–5012. [CrossRef]
46. Bussolati, B.; Collino, F.; Camussi, G. CD133+ cells as a therapeutic target for kidney diseases. *Expert. Opin. Ther. Targets* **2012**, *16*, 157–165. [CrossRef]
47. Wang, H.; Wang, X.; Xu, L.; Zhang, J.; Cao, H. A pan-cancer perspective analysis reveals the opposite prognostic significance of CD133 in lower grade glioma and papillary renal cell carcinoma. *Sci. Prog.* **2021**, *104*, 2021. [CrossRef]
48. Beck, A.C.; Rajan, A.; Landers, S.; Kelley, S.; Bellizzi, A.M.; Lal, G.; Sugg, S.L.; Howe, J.R.; Chan, C.H.; Weigel, R.J. Expression of cancer stem cell markers in tall cell variant papillary thyroid cancer identifies a molecular profile predictive of recurrence in classic papillary thyroid cancer. *Surgery* **2022**, *171*, 245–251. [CrossRef]
49. Saeednejad Zanjani, L.; Madjd, Z.; Abolhasani, M.; Andersson, Y.; Rasti, A.; Shariftabrizi, A.; Asgari, M. Cytoplasmic expression of CD133 stemness marker is associated with tumor aggressiveness in clear cell renal cell carcinoma. *Exp. Mol. Pathol.* **2017**, *103*, 218–228. [CrossRef]
50. Hasmim, M.; Bruno, S.; Azzi, S.; Gallerne, C.; Michel, J.G.; Chiabotto, G.; Lecoz, V.; Romei, C.; Spaggiari, G.M.; Pezzolo, A.; et al. Isolation and characterization of renal cancer stem cells from patient-derived xenografts. *Oncotarget* **2015**, *7*, 15507. [CrossRef]
51. Feng, G.; Jiang, F.; Pan, C.; Pu, C.; Huang, H.; Li, G. Quantification of peripheral blood CD133 mRNA in identifying metastasis and in predicting recurrence of patients with clear cell renal cell carcinoma. *Urol. Oncol. Semin. Orig. Investig.* **2014**, *32*, 44.e9–44.e14. [CrossRef]
52. Bradley, J.R.; Wang, J.; Pacey, S.; Warren, A.Y.; Pober, J.S.; Al-Lamki, R.S. Tumor necrosis factor receptor-2 signaling pathways promote survival of cancer stem-like CD133+ cells in clear cell renal carcinoma. *FASEB Bioadv.* **2020**, *2*, 126. [CrossRef] [PubMed]
53. Ferrari, I.; De Grossi, F.; Lai, G.; Oliveto, S.; Deroma, G.; Biffo, S.; Manfrini, N. CancerHubs: A systematic data mining and elaboration approach for identifying novel cancer-related protein interaction hubs. *Brief. Bioinform.* **2024**, *26*, bbae635. [CrossRef] [PubMed]
54. Graham, J.; Heng, D.Y.C.; Brugarolas, J.; Vaishampayan, U. Personalized Management of Advanced Kidney Cancer. *Am Soc Clin Oncol Educ Book* **2018**, *38*, 330. [CrossRef] [PubMed]
55. Saliby, R.M.; Saad, E.; Kashima, S.; Schoenfeld, D.A.; Braun, D.A. Update on Biomarkers in Renal Cell Carcinoma. *Am. Soc. Clin. Oncol. Educ. Book* **2024**, *44*, e430734. [CrossRef]
56. Laru, L.; Ronkainen, H.; Ohtonen, P.; Vaarala, M.H. The impact of metastasectomy on survival of patients with synchronous metastatic renal cell cancer in Finland: A nationwide study. *Scand. J. Surg.* **2024**, *113*, 219–228. [CrossRef]
57. Ferriero, M.; Cacciatore, L.; Ochoa, M.; Mastroianni, R.; Tuderti, G.; Costantini, M.; Anceschi, U.; Misuraca, L.; Brassetti, A.; Guaglianone, S.; et al. The Impact of Metastasectomy on Survival Outcomes of Renal Cell Carcinoma: A 10-Year Single Center Experience. *Cancers* **2023**, *15*, 3332. [CrossRef]
58. Oki, R.; Takemura, K.; Urasaki, T.; Fujiwara, R.; Numao, N.; Yonese, J.; Miura, Y.; Yuasa, T. Prevailing challenges in personalized treatment for metastatic renal cell carcinoma: A narrative review. *Expert Rev. Anticancer Ther.* **2025**, *25*, 643–655. [CrossRef]

MDPI AG
Grosspeteranlage 5
4052 Basel
Switzerland
Tel.: +41 61 683 77 34

International Journal of Molecular Sciences Editorial Office
E-mail: ijms@mdpi.com
www.mdpi.com/journal/ijms

www.ingramcontent.com/pod-product-compliance
Lightning Source LLC
Chambersburg PA
CBHW062103220926
51850CB00010B/2866
* 9 7 8 3 7 2 5 8 6 6 4 6 5 *